·uantitative Bioassay

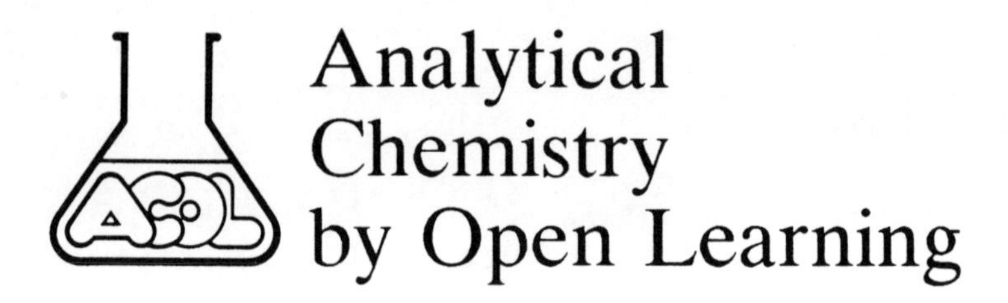
Analytical
Chemistry
by Open Learning

**Titles in Series:**

Samples and Standards
Sample Pretreatment and Separation
Classical Methods
Measurement, Statistics and Computation
Using Literature
Instrumentation
Chromatographic Separations
Gas Chromatography
High Performance Liquid Chromatography
Electrophoresis
Thin Layer Chromatography
Visible and Ultraviolet Spectroscopy
Fluorescence and Phosphorescence Spectroscopy
Infra Red Spectroscopy
Atomic Absorption and Emission Spectroscopy
Nuclear Magnetic Resonance Spectroscopy
X-ray Methods
Mass Spectrometry
Scanning Electron Microscopy and X-Ray Microanalysis
Principles of Electroanalytical Methods
Potentiometry and Ion Selective Electrodes
Polarography and Other Voltammetric Methods
Radiochemical Methods
Clinical Specimens
Diagnostic Enzymology
Quantitative Bioassay
Assessment and Control of Biochemical Methods
Thermal Methods
Microprocessor Applications

# Quantitative Bioassay

Analytical Chemistry by Open Learning

Authors:
DAVID HAWCROFT and TERRY HECTOR
*Leicester Polytechnic*

FRED ROWELL
*Sunderland Polytechnic*

Editor:
ARTHUR M. JAMES

*on behalf of ACOL*

Published on behalf of ACOL, Thames Polytechnic, London
by
JOHN WILEY & SONS
Chichester · New York · Brisbane · Toronto · Singapore

***Library of Congress Cataloging in Publication Data:***

Hawcroft, David M.
Quantitative bioassay / authors, David Hawcroft and Terry Hector, Fred Rowell ; editor, Arthur M. James.
p. cm. — (Analytical Chemistry by Open Learning)
ISBN 0 471 91400 2 : ISBN 0 471 91401 0 (pbk.) :
1. Biomolecules—Analysis. 2. Biological assay.
3. Microbiological assay. 4. Immunoassay. 5. Chemistry, Analytic—Quantitative. I. Hector, Terry. II. Rowell, Fred. III. James, A. M. (Arthur M.), 1923– . IV. ACOL (Project) V. Title.
VI. Series: Analytical Chemistry by Open Learning (Series)
QP519.7.H39 1988
574.19'285—dc19 87-36006
CIP

***British Library Cataloguing in Publication Data:***

Hawcroft, David
Quantitative bioassay.
1. Quantitative biological assay.
I. Title II. Hector, Terry III. Rowell,
Fred IV. James, Arthur M., 1923–
V. ACOL VI. Series
615'.19018

ISBN 0 471 91400 2
ISBN 0 471 91401 0 Pbk

Printed and bound in Great Britain

# Analytical Chemistry

This series of texts is a result of an initiative by the Committee of Heads of Polytechnic Chemistry Departments in the United Kingdom. A project team based at Thames Polytechnic using funds available from the Manpower Services Commission 'Open Tech' Project has organised and managed the development of the material suitable for use by 'Distance Learners'. The contents of the various units have been identified, planned and written almost exclusively by groups of polytechnic staff, who are both expert in the subject area and are currently teaching in analytical chemistry.

The texts are for those interested in the basics of analytical chemistry and instrumental techniques who wish to study in a more flexible way than traditional institute attendance or to augment such attendance. A series of these units may be used by those undertaking courses leading to BTEC (levels IV and V), Royal Society of Chemistry (Certificates of Applied Chemistry) or other qualifications. The level is thus that of Senior Technician.

It is emphasised however that whilst the theoretical aspects of analytical chemistry can be studied in this way there is no substitute for the laboratory to learn the associated practical skills. In the U.K. there are nominated Polytechnics, Colleges and other Institutions who offer tutorial and practical support to achieve the practical objectives identified within each text. It is expected that many institutions worldwide will also provide such support.

The project will continue at Thames Polytechnic to support these 'Open Learning Texts', to continually refresh and update the material and to extend its coverage.

Further information about nominated support centres, the material or open learning techniques may be obtained from the project office at Thames Polytechnic, ACOL, Wellington St., Woolwich, London, SE18 6PF.

# How to Use an Open Learning Text

Open learning texts are designed as a convenient and flexible way of studying for people who, for a variety of reasons cannot use conventional education courses. You will learn from this text the principles of one subject in Analytical Chemistry, but only by putting this knowledge into practice, under professional supervision, will you gain a full understanding of the analytical techniques described.

To achieve the full benefit from an open learning text you need to plan your place and time of study.

- Find the most suitable place to study where you can work without disturbance.

- If you have a tutor supervising your study discuss with him, or her, the date by which you should have completed this text.

- Some people study perfectly well in irregular bursts, however most students find that setting aside a certain number of hours each day is the most satisfactory method. It is for you to decide which pattern of study suits you best.

- If you decide to study for several hours at once, take short breaks of five or ten minutes every half hour or so. You will find that this method maintains a higher overall level of concentration.

Before you begin a detailed reading of the text, familiarise yourself with the general layout of the material. Have a look at the course contents list at the front of the book and flip through the pages to get a general impression of the way the subject is dealt with. You will find that there is space on the pages to make comments alongside the text as you study—your own notes for highlighting points that you feel are particularly important. Indicate in the margin the points you would like to discuss further with a tutor or fellow student. When you come to revise, these personal study notes will be very useful.

Π When you find a paragraph in the text marked with a symbol such as is shown here, this is where you get involved. At this point you are directed to do things: draw graphs, answer questions, perform calculations, etc. Do make an attempt at these activities. If necessary cover the succeeding response with a piece of paper until you are ready to read on. This is an opportunity for you to learn by participating in the subject and although the text continues by discussing your response, there is no better way to learn than by working things out for yourself.

We have introduced self assessment questions (SAQ) at appropriate places in the text. These SAQs provide for you a way of finding out if you understand what you have just been studying. There is space on the page for your answer and for any comments you want to add after reading the author's response. You will find the author's response to each SAQ at the end of the text. Compare what you have written with the response provided and read the discussion and advice.

At intervals in the text you will find a Summary and List of Objectives. The Summary will emphasise the important points covered by the material you have just read and the Objectives will give you a checklist of tasks you should then be able to achieve.

You can revise the Unit, perhaps for a formal examination, by rereading the Summary and the Objectives, and by working through some of the SAQs. This should quickly alert you to areas of the text that need further study.

At the end of the book you will find for reference lists of commonly used scientific symbols and values, units of measurement and also a periodic table.

# Contents

# Study Guide

This Unit deals with three principal types of assay procedure, namely those using animals or their tissues, assays using microorganisms and immunoassays.

Parts 1 and 2 of the Unit are concerned with assays that use animals and microorganisms respectively. That these two parts together constitute less than half of the total Unit is appropriate and reflects the declining importance of the classical bioassays as they are replaced by an ever-growing range of immunoassay techniques.

There is considerable and justified pressure from society to reduce and eventually eliminate any mental and physical suffering of animals in laboratories. This alone is sufficient stimulus for the scientific community to strive for alternatives to the use of animals in research and development establishments. Most developed nations still allow animals to be used by laboratories but only under strictly controlled conditions, and in Britain all laboratories are subject to regular inspections. Demanding criteria are laid down for the care of laboratory animals and the facilities which must be provided for them, and these make animal work extremely expensive. When, finally, the relatively poor analytical performances of bioassays using animals are considered it is manifestly obvious that they should be regarded as 'last resort' methods, ie only undertaken when no valid alternatives exist.

Microbiological assays can be regarded more favourably than assays involving animals. There is no moral debate that we are aware of concerning the rights and wrongs of experimentation with bacteria, fungi, protozoa, etc. The genetic variations of individuals within a pure culture of microorganisms such as bacteria are negligible, unlike differences which can be demonstrated even within in-bred strains of mice, and the relatively low cost of microbiological assay techniques allows for considerable replication of tests. Thus the analytical performance of a microbiological assay is generally far better than that of a bioassay using animals or animal tissues. Microbiological assays are limited however to measuring substances which either inhibit or promote the growth of a microorganism.

Immunoassays involve the use of antibodies, these being proteins produced by the body in response to the presence of foreign (ie not of that body) substances, and which are extremely useful because they can bind other compounds with a degree of specificity. Complete kits for established immunological methods are available for the identification and measurement of a very wide range of substances of biological, including medical importance. Although the kits are relatively expensive, these methods are extensively and increasingly used in clinical biochemistry, forensic, toxicology and other pathology and indeed non-pathology, laboratories. This is because they provide a simplicity, specificity, and sensitivity (frequently $10^{-12}$ mol $dm^{-3}$ or better), which few other analytical methods can match, especially for important but unusual groups of compounds such as hormones and vitamins.

While these kits allow someone with virtually no knowledge of biology to carry out the measurements, an understanding of the basics of protein chemistry and immunology will, apart from being of some academic and educational interest, increase your awareness of the principles of the methods and give you a feeling for their potential and limitations. As a result of understanding these basics you might also be able to consider devising your own assay for a particular application.

Consequently Part 3 begins with a consideration of the nature of antigenic compounds and their antibodies, their preparation and those properties relevant to the techniques in question. Later parts of the Unit consider those assay methods that are most significant in quantitative analytical biochemistry. It is not appropriate in a text of this type to attempt an exhaustive coverage of all the variants of the basic procedures and in particular we have restricted ourselves to those in which the antibody acts as a reagent rather than the target of the assay, and furthermore a consideration is given only to the major sub-divisions of the techniques. More details of all these techniques can be obtained from the texts given in the Bibliography.

Immunological principles are involved in many other types of analysis, not all of which are quantitative, and a brief mention of these is included for the sake of completeness. The Glossary provided towards the end of the Unit contains definitions of many immunological terms although they are defined when first used in the text. Such a collection is likely to be useful because of the considerable number of specialised terms encountered in this subject.

If you have a limited knowledge of protein chemistry and immunology you should treat Parts 3 to 6 as a single entity and begin your study with Part 3. If you have some basic knowledge of these areas you could perhaps start with a later Part, perhaps using the questions of previous Parts as revision or test exercises.

# Supporting Practical Work

### Practical work involving animals

In view of the considerable opposition to the use of animals in teaching situations this type of work is best restricted to colleges which have the necessary resources and staff with adequate skills and experience. Many experiments and assay techniques are possible, but which of these are acceptable is a matter for discussion and agreement between senior college staff and representatives of the national licensing authority, ie The Home Office in Britain.

### Practical work involving microorganisms

It is much easier to identify and carry out suitable work using microorganisms of which the following list is a representative selection.

1. Identification of the sensitivity of selected bacteria to a range of different antibiotics using commercial paper disc products.

2. Determination of the minimum inhibitory concentration of an antibiotic (eg penicillin) on a reference strain of *Staphylococcus aureus* by a tube dilution technique.

3. Determination of the minimum inhibitory concentration of an antibiotic (eg penicillin) on test organisms using a disc diffusion technique.

4. The effect of protein concentration on penicillin activity in plate assays.

5. The effect of thickness of medium, size of inocula and pre-incubation periods on zone diameters in plate assays.

6. The determination of vitamin $B_{12}$ in plasma using a tube technique.

**Practical work on immunoassays**

The selection of appropriate experimental work to support the immunoassay part of this Unit presents some problems. The raising of antibodies is not really a suitable exercise since it requires very specialised facilities, staff with the appropriate Home Office licences and is the subject of specific and rigorous legal control. While the generation and purification of an antiserum is an interesting area it is somewhat similar to the synthesis and purification of a chemical reagent before its use in a more traditional area of chemical assay, and as such is not very appropriate. The properties of the antiserum are key factors in the performance of an immunoassay and it is certainly appropriate to study these. However, for those readers who are interested in the purification of antisera relevant exercises are included in some of the texts listed in the experimental immunology section of the Bibliography, especially perhaps Hudson and Hay (1980). These texts also contain details of many simple experiments using other important immunological techniques. With regard to the immunoassay procedures discussed in this Unit, it is the case that many laboratories make extensive use of purchased reagents and complete assay kits and using these, procedures frequently become very straightforward and of little educational value. In contrast the development and evaluation of an immunoassay from the basic components can be very time consuming, expensive and demand reasonable technical skill. There are also important safety and legal considerations if a radio-isotope label is to be used.

However experience has shown that it is possible to develop either from first principles if time is available (say as a project), or using some prepared reagents (eg a purchased labelled antigen), one of the more straightforward enzyme-labelled, or fluorescence-labelled immunoassay systems. To do this, it is recommended that reference is made to the texts listed in the Bibliography and through them, or independently, to specific papers in the scientific literature.

# Bibliography


**Macrobiological Assay and Experimental Design**

Finney, D J. *Statistical Method in Biological Assay*. 3rd ed. Griffin, London, 1978.

Johnson, F N and Johnson, S. *Clinical Trials*. Blackwell, Oxford, 1977.

Whitehead, J. *The Design and Analysis of Sequential Clinical Trials*. Ellis Horwood, Chichester, 1983.

Rowan, A N and Stratmann, C J. *The Use of Alternatives in Drug Research*. Macmillan, London, 1980.

Heath, O V S. *Investigation by Experiment*. Arnold, London, 1970.

Miller, S. *Experimental Design and Statistics*. 2nd ed, Methuen, London, 1984.

**Basic Microbiological technique**

Kavanagh, F W. *Analytical Microbiology* Vol I and II. Academic Press, 1963 and 1971.

Hewitt, W. *Microbiological Assay*. Academic Press, 1977.

*The National Collection of Industrial Bacteria Catalogue of Strains*. H.M.S.O. 3rd ed, 1975.

*Difco Manual*, 10 ed, Difco Laboratories, 1987.

Strohecker, R and Henning, H M. *Vitamin Assay*. Verlage Chemie, 2nd ed, 1972.

Collins, C H and Lyne, P M. chapters 1-5. *Microbiological Methods*. Butterworths. 5th ed, 1984.

**Basic Immunology**

Roitt, I., Brostoff, J. and Male, D. *Immunology*. Livingstone, Edinburgh, 1985.

Roitt, I. *Essential Immunology* 5th ed. Blackwell, Oxford, 1984.

Bier, O., Da Silva, W., Gotze, D., and Mota, I. *Fundamentals of Immunology* 2nd ed. Springer Verlag, Berlin, 1986.

**Experimental Immunology and Immunoassay**

Barnard, G J R and Collins, W P. *The Development of Luminescence Immunoassays*. Medical Laboratory Sciences 44, 249-266, 1987.

Butt, W. (Ed). *Practical Immunology*. Marcel Dekker, New York, 1984.

Collins, W. (Ed). *Alternative Immunoassays*. Wiley, Chichester, 1985.

Edwards, R. *Immunoassay, an Introduction*. Heinemann, London, 1985.

Hunter, W and Corrie, J. *Immunoassay in Clinical Chemistry*, 2nd ed. Churchill, 1983.

Maggio, E. *Enzyme Immunoassay*. CRC Press, Boca Raton, Florida, 1980.

Weir, D. (Ed). *Handbook of Experimental Immunology*, 3rd ed. Blackwell, Oxford, 1978.

Hudson, L and Hay, F. *Practical Immunology*, 2nd ed. Blackwell, Oxford, 1980.

Chard, T. *An Introduction to Radioimmunoassay and Related Techniques*, in *Laboratory Techniques in Biochemistry and Molecular Biology, Vol 6*, p293-534. Work T and Work E (Eds). Elsevier, 1978.

# Acknowledgement

The ACOL project team would like to thank Dr Richard Solly of Newcastle Polytechnic for his assistance in the preparation of ACOL: Quantitative Bioassay.

# 1. General Aspects of Bioassays

## Overview

The general principles of bioassays are introduced in this Part with particular attention to those assays which use animals and their tissues. The advantages and disadvantages associated with these assays are discussed with reference to both the scientific considerations and the moral issues that the use of animals raises. The problems of biological variation between individuals and the necessity for careful experimental design are explained.

## 1.1. BASIC PRINCIPLES

Bioassays involve quantitation of the response which follows the application of a stimulus to a biological system, ie.

STIMULUS + BIOLOGICAL SYSTEM → RESPONSE

The applied stimulus is represented by standard or test samples which contain the biologically active substance or analyte. The biological system which receives the stimulus may be a whole, multicellular organism (animal or plant), isolated organs or tissues from multicellular organisms, whole cells (microorganisms) or biologically active macromolecules (antibodies, enzymes) that are produced by living organisms. The response is a change seen in some aspect of the biological system. It may be a positive response associated with increased activity or a negative response that is inhibitory or even lethal to the biological system.

When living organisms, tissues or cells are used in bioassays the response relates to a biological activity that is normally attributed to the analyte. For example the drug digitalis can be measured by its effect on the contractions of isolated muscle, or the activity of an antibiotic such as penicillin determined by its ability to inhibit the growth of a susceptible bacterium in an artificial culture medium. This Part and Part 2 of the text refer mainly to these types of bioassays. The remaining text is devoted principally to immunoassays. An immunoassay depends on a reaction between the analyte in standard or test samples and antibodies which have been produced by cells of an animal's immune system. The ease with which immunoassays are performed, their good analytical performance and economy provide ample justification for their current popularity over other bioassay methods. It is important however to recognise that the results given by an immunoassay depend on the specificity of the antibody for the biologically-active analyte in standard and test materials. Misleading results have been given by the use of antibodies that react with biologically-inactive precursors or metabolic derivatives of the active analyte.

## 1.2. THE USE OF ANIMALS AND ANIMAL TISSUES

The use of animals or their tissues is an emotive subject that raises moral issues for both the scientific community and the general public. Experiments that involve animals have been strictly controlled in Britain by the Home Office, previously under legislation contained in the 'Cruelty to Animals Act, 1876', and currently by the 'Animals Scientific Procedures Act, 1986'. For both scientific and humane reasons the use of animals should be considered only at a last resort.

The necessity to use animals for bioassays has declined over the years and continues to do so. Public opinion and organised pressure groups such as FRAME (Fund for Replacement of Animals in Medical Research) in Britain have undoubtably influenced this trend. At one time all new compounds considered by the pharmaceutical companies to be potential drugs were screened against vast numbers of laboratory animals. It is now possible to perform initial screening tests using techniques that involve isolated cells maintained in artificial culture media. These techniques allow all those compounds which cause death or obvious morphological or biochemical changes to the cells to be eliminated as potential drugs and restricts tests on animals to a minimum.

Even in carefully designed bioassays using animals or their tissues the precision is often poor and the procedures are generally costly. Therefore, quite aside from any humane considerations, the scientific justification to employ these types of bioassays can be made only when no suitable alternative assay methods are available. If the term bioassay is extended to include the clinical trials of newly developed drugs on selected human populations, the problems of ethics (at one time prison inmates were put under some pressure to partake in these trials) and of experimental design become very important indeed.

One advantage of bioassays is that they are generally specific for biologically-active forms of the analyte. Good specificity may make a bioassay the method of choice when the test material contains a mixture of active and inactive forms of the analyte which cannot be separated effectively. Alternatively the analyte may occur in a variety of active forms which display different biological activities and are present in unknown relative quantities. In both cases, a physical or chemical assay may measure the total quantity of analyte, but in doing so give no indication of the biological activity of the sample. This type of situation could arise with material that has been extracted from an endocrine gland such as the pituitary. The extract may contain several hormones which affect the same target cells but with different relative biological activities.

In addition to high specificity, bioassays are often very sensitive. In this context sensitivity indicates the ability of the assay to distinguish between small differences in the analyte activities of samples, which in turn means that low levels of activity can be measured. It may well be the case in a research project that there is insufficient material available for purification procedures to be attempted. In such situations the high sensitivity of bioassays can give them a distinct advantage over alternative methods. It must be said however that this reason for performing bioassays which use animals or their tissues is becoming less frequent as purification and alternative analytical techniques continue to improve.

**SAQ 1.2b**

Select from the following options valid reasons for using bioassays involving animals as a last resort.

(*i*) They may be expensive and time consuming to perform.

(*ii*) They can only be used for biologically-active compounds.

(*iii*) They generally give poor precision.

(*iv*) In some types of assays animals may die or have to be slaughtered.

## 1.3. BIOLOGICAL ACTIVITY AND STANDARD MATERIALS

Biological activity or potency refers to the quantitative effect which a sample preparation has on a biological system. There is a tendency to use concentration units as a measure of potency in bioassays. This is acceptable if we are always aware that it is not an absolute concentration which is being measured, but one which is relative to an arbitrary, albeit internationally recognised, standard. Also we can never be quite sure whether the assay is giving a measure of the activity of a single compound or a mixture of compounds with common biological activities.

All bioassays are comparative and require a standard preparation with which each test sample can be compared. The general principle which is applied in bioassays is that like is compared with like. This means that the standard preparation must owe its biological activity to the same active substances that are present in the test sample. For example, if we are performing an assay on an extract of the posterior pituitary gland, then the standard preparation should have been derived from that same source material. Strictly speaking, the total composition of the test and standards should be identical although this is virtually impossible to achieve in practice.

At the local level, in perhaps a research project where a novel assay has been developed, the scientists involved may have to produce their own standard. All test sample responses would be compared with the response of this standard preparation to which arbitrary units of activity are assigned.

When an assay is performed in several establishments the policy of using 'in-house' standards is not valid and can only result in a lack of inter-laboratory correlation. Correlation between laboratories is achieved by the use of International Standards for the Calibration of Bioassays. The provision of International Standards for bioassays is one of the responsibilities of the World Health Organisation who are advised by the Expert Committee on Biological Standardisation. Standards for immunological substances are made, held and distributed by the Department of Biological Standards, Statens Seruminstitut in Copenhagen, Denmark, and those for nonimmunological substances by the National Institute for Medical Research, London, United Kingdom. An International Standard must be produced in quantities that are sufficiently large to meet world demand for several years. The standard material must be stored in sealed containers protected from the harmful effects of light, moisture and air and maintained at a constant temperature. It will come as no surprise to learn that International Standards are very expensive. National laboratories will generally hold working standards which have been calibrated against the appropriate International Standards.

The standard for a bioassay will have been assigned arbitrary units of activity. For example, in its account of the bioassay for insulin the British Pharmacopeia describes the standard as follows:

'The Fourth International Standard, established in 1958, consisting of a mixture of purified, recrystallised beef and pork insulin (containing 24.0 Units per mg), or another suitable preparation, the potency of which has been determined in relation to the International Standard'.

Each standard material has its own unit of activity which is expressed as an arbitrary weight in milligrams of the current International Standard. The relationship between International Units of potency and the weight of one substance is distinct from that of other substances. There is also no relationship between the therapeutic number of units required of one substance compared with another. For example, one Unit of insulin is 0.04167 mg of the International Standard and the therapeutic dose of insulin lies in the range 10 to 100 Units. This contrasts with one Unit of chorionic gonadotrophin which is 0.1 mg of the International Standard and which has a therapeutic dose range of 500 to 1000 Units.

**SAQ 1.3a**

Indicate which of the following statements about bioassays are correct.

(*i*) Bioassays can be applied to any analyte.

(*ii*) Pure analytical grade standard is required in bioassays.

(*iii*) Bioassays measure the biological response of a living organism to the analyte.

(*iv*) The greater the biological response of the organism to the analyte the greater the potency of the analyte.

(*v*) The potency and biological activity of the analyte are synonymous.

(*vi*) The potencies of analytes in bioassays are expressed in mass units of mass per volume.

## 1.4. TYPES OF EFFECTS

### 1.4.1. The use of live animals

A bioassay performed on living animals generally depends on the observation of a particular physiological response. For example, it is possible to assess the activity of a product containing vitamin D by its ability to prevent the development of rickets in rats fed on vitamin D deficient diet. Immunological activity has been assessed by the ability of specific antibodies to neutralise the physiological effects in laboratory animals of antigenic material eg exotoxins of the fatal food poisoning bacterium *Clostridium botulinum* in a test sample. Assays have even been carried out on human volunteers. The irritant activity of biocides have been assessed by the 'arm dip test'. This test consists of groups of human volunteers immersing their arms into test solutions of biocides at various concentrations for 15 minutes, twice a day for up to 4 weeks. The skin is examined during this time for signs of irritation.

### 1.4.2. The use of isolated tissues

Certain tissues are the sites at which the biological response takes place in the animal. We can use these tissues provided that they are freshly removed from the animal so that they will still respond to the active constituents within the sample.

A response that involves an isolated tissue rather than the whole animal is that used for the assay of histamine. You will be familiar with the outwardly visible effects (runny nose and eyes, swelling etc) that histamine released from body cells has on people who suffer from hay fever. Among the many other effects of histamine is its influence on the contraction of smooth muscle and this effect is used in its assay. The assay employs short lengths of guinea-pig small intestine. Contractions of small intestine on exposure to test samples are compared with contractions of the tissue on exposure to standards with known histamine activities.

Responses given by methods such as the histamine assay are easily measured in recognised units (cm, sec, etc) and the response changes in proportion to the strength of the stimulus. However there are responses which are not continuously variable according to the intensity of stimulation. All that can be said about certain responses (death or pregnancy) is that they have or have not occurred, it is not possible to have 75% death or to be slightly pregnant. These are all-or-nothing or quantal responses. In a procedure that gives a quantal response, it is possible to obtain a scaled result by performing the method many times and calculating the frequency of the effect, eg the percentage of animals which died when exposed to a particular toxin dose. For statistical analysis results are often converted into probits (probability units) using statistical tables.

**SAQ 1.4a**

Which of the following could be classed as procedures that give quantal responses?

(*i*) Doses of weedkiller which cause the death of rats.

(*ii*) Doses of toxin which produce visible convulsions in the hind legs of guinea-pigs.

(*iii*) Doses of prothrombin on the time required for fibrin clots to form in plasma samples.

(*iv*) Doses of radiation on the life span of selected laboratory animals.

**SAQ 1.4a**

Having obtained a clearly measurable response, either directly or by conversion of a set of quantal responses into a percentage or probit expression, its relationship to the intensity of the stimulus must be established. The ideal relationship which is both easy to plot and analyse is direct and rectilinear (ie a straight line plot). A simple rectilinear response is sadly rare but the log dose/response occurs commonly. Other relationships also exist such as the log dose/log response or the log dose-logit plot used in immunoassays and described in Section 6.5 of the text. Often we see a direct relationship between the two parameters but inverse relationships are by no means uncommon. For example an increase in the levels of blood prothrombin reduces the time required for fibrin clots to form.

The slope of the line and its position on the *x*-axis are important since they show the sensitivity and detection limits (range) of the assay respectively. The response to an applied stimulus should be sufficiently sensitive to be able to differentiate between small differences in doses but not so great that it restricts the detection limits of the assay. Fig. 1.4a, demonstrates various responses in graphic form.

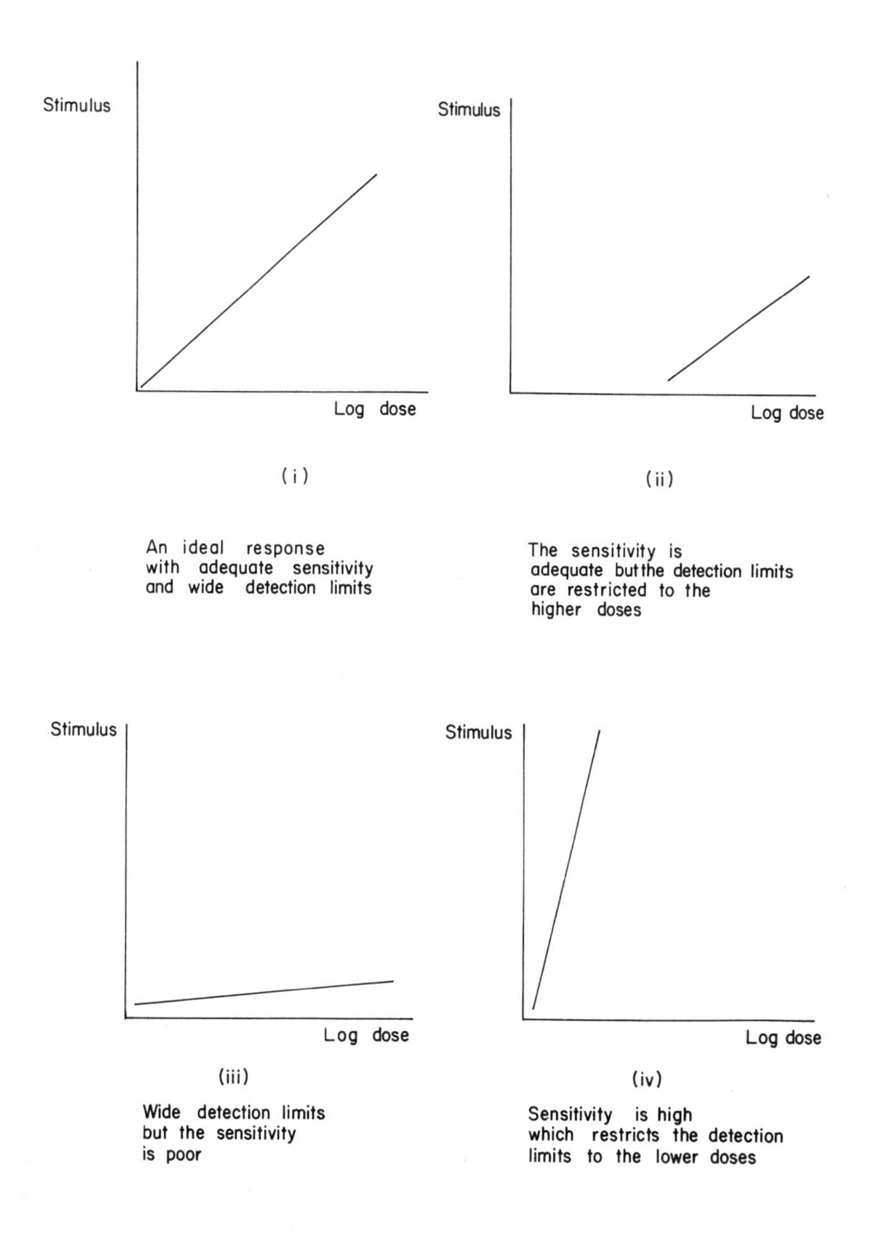


**Fig. 1.4a.** *Various stimulus/log dose responses in bioassays*

## 1.5. EXPERIMENTAL DESIGN

All experimental procedures benefit from repetition and bioassays are no exception. In fact the problem of variability of response in most types of bioassays is so great that repetition is essential.

Consider an assay in which the response of one animal to a test dose of a drug is compared with the response of a second animal who receives a standard dose of the drug. This assay procedure could be repeated every day for two weeks and provide a plentiful supply of data from which the mean values for both the test and the standard responses could be calculated. It is evident that the designer of this assay appreciated the need to repeat his measurements but let us examine whether repetition alone is sufficient. Firstly, the assay is very time consuming and this may introduce temporal variations in the responses. Temporal variations (variations associated with time) are caused by the biochemical, physiological and psychological changes that take place as the animal reacts to its environment during the period of the test. Secondly, the assay design makes no allowance for biological variations. Good precision as reflected by close agreement between the replicate doses may be achieved, but a large degree of bias (inaccuracy) due to differences in the responses of the two animals would go undetected.

In the assay described above it would have been possible to correct for biological differences in the animals by switching the test and standard doses between the two of them. The order in which the doses were given could also have been arranged to allow for temporal variations and for the effects of one dose on the next.

Π You are asked to decide how the treatments might be ordered in a simple assay procedure on a piece of isolated tissue.

Imagine a bioassay in which both the standard dose (S) and test dose (T) can be applied to the same piece of isolated tissue, ie two S and two T in any sequence. Which of the following sequences of treatments do you consider to be most capable of revealing temporal effects including the effects of standard dose on test doses and *vice versa*?

(*i*) SSTT

(*ii*) TTSS

(*iii*) STST

(*iv*) STTS

STTS,(*iv*), is probably the best choice since it does at least allow for the influence of the standard dose on the following test dose and *vice versa.*

Thus we see that experimental design involves more than mere repetition. Randomisation is required, and in this case it has been achieved to a limited degree by using an approach called the cross-over technique. In practice several groups of animals or tissues would be used to give a measure of replication. It is also likely that more than one dose of both the test and standard materials would be employed. SAQ 1.5a illustrates a more practical example of a cross-over assay.

**SAQ 1.5a**

Four groups of animals were subject to a total of four doses of drug comprising high and low standard doses (SH and SL) and high and low test doses (TH and TL). A cross-over assay was then performed using the following pattern:

| | First exposure | Second exposure |
|---|---|---|
| Group A | SH | TH |
| Group B | SL | TL |
| Group C | TH | SH |
| Group D | TL | SL |

How could randomisation of the above pattern be improved?

**SAQ 1.5a**

A popular means of obtaining randomisation combined with replication is the use the Latin square design. In a true Latin square each treatment appears once in each column and once in each row. The assay which was described in SAQ 1.5a could have been expanded to the Latin.square design shown in Fig. 1.5a. Each group of animals receive the two test and two standard doses in sequence over a four day period.

| Groups of animals | Days 1 | 2 | 3 | 4 |
|---|---|---|---|---|
| A | SL | TL | TH | SH |
| B | SH | TH | TL | SL |
| C | TH | SH | SL | TL |
| D | TL | SL | SH | TH |

**Fig. 1.5a.** *A Latin square design in which each group of animals receives a sequence of two test and two standard treatments*

Imagine that the workload of the assay just described is such that any one laboratory worker can only manage one group of animals each day. Four laboratory workers must therefore be employed each perhaps differing slightly in their technique and perhaps improving their individual performances over the course of the assay (ie they become more skillful by day 4). It is possible to take this additional factor into the design by using the more complex Graeco-Latin square. The number of different Graeco-Latin squares is severely limited compared with Latin square of the same size. Fig. 1.5b shows how the Latin square design shown in Fig. 1.5a can be modified to produce a Graeco-Latin square. The four different laboratory workers are identified by the appropriate use of Greek letters each of which appears once in each column and once in each row.

Latin square designs and the construction of balanced assays (2+2, 3+3 etc) are discussed further in Part 2 of the text.

| Groups of animals | Days 1 | 2 | 3 | 4 |
|---|---|---|---|---|
| A | SL$\gamma$ | TL$\delta$ | TH$\alpha$ | SH$\beta$ |
| B | SH$\delta$ | TH$\gamma$ | TL$\beta$ | SL$\alpha$ |
| C | TH$\beta$ | SH$\alpha$ | SL$\delta$ | TL$\gamma$ |
| D | TL$\alpha$ | SL$\beta$ | SH$\gamma$ | TH$\delta$ |

**Fig. 1.5b.** *A Graeco-Latin square design in which four groups of animals receive test and standard treatments over four days administered by four different laboratory workers ($\alpha$, $\beta$, $\gamma$ and $\delta$)*

While the examples which have just been described required several animals, it is generally possible to use each animal for a sequence of treatments in order to minimise the effects of biological (or spatial) variability. However with quantal assays this is not possible, and ideally a large number of animals should be used. This presents problems. We must attempt to ensure that there is minimal variation between the organisms used, so that they all exhibit the same response when exposed to the same test dose. In microbiological assays it is possible to generate cultures of microorganisms containing millions of genetically identical cells which should therefore, under constant conditions, respond in the same way. Although genetic identity can be ensured with cultures of isolated animal cells, in general, this is not possible with bioassays using whole animals or their tissues.

The usual approach therefore is to select groups of animals which are matched as closely as possible with regard to their gender, weight, age and any other factors which may be relevant to the study. In particular it is useful if the animals can come from the same genetic line, perhaps even as siblings, and many laboratories which routinely carry out this work maintain in-bred colonies of test animals in order to minimise genetic variability. In addition, once the conditions for an assay have been optimised it is essential that they are defined so that intra- and inter-laboratory comparisons of assay results may be made. These conditions include the pretreatment of animals (specific diets, starvation, surgery), compositions of bathing fluids for isolated tissues and, when whole animals are used, the housing conditions should be stated. Even the litter used can influence results, for example sawdust contains phenols which affect liver metabolism.

The combination of a stated biological system and controlled experimental conditions together constitute the 'restriction in design' of the assay. The more stringent the restriction the more reliable is the assay, but this inevitably means greater expense all round in terms of increased time, organisms and apparatus. The following is an extract from a bioassay method for insulin. It demonstrates the level of restriction in design that is demanded when performing official standard tests. 'The mice should not vary by more than 5 g in weight and must have been fed on an adequate diet and then deprived of food for not less than 2, or more than 20 hours, before the assay. After injection with the samples of insulin they must be kept in an incubator at a constant temperature between 29 and 35 °C. This incubator may be a glass-fronted air incubator with the mice kept in glass jars, or a series of small boxes two-thirds immersed in a water bath maintained at the appropriate temperature'.

Selection of the number of animals to use in an assay is a very complex matter in which statistical considerations are of the utmost importance. In principle it is necessary to choose sufficient numbers of animals within each group so that the response which any one individual has on the mean response is minimised. However because we are using animals an arbitrarily determined large number 'to be on the safe side' cannot be justified for ethical reasons and would, in any case, be extremely expensive (the cost of keeping a laboratory rat for 3–4 years is several hundred pounds). It is necessary to carry out preliminary studies to predict the likely biological variability and to make an estimate of the minimum numbers of animals for the study. It is no wonder that most institutions insist that experimental designs for this type of work must be approved by statisticians before being undertaken.

Having determined the number of animals that are required, it is important that they are used to maximum effect which also involves statistical aspects of experimental design. The animals need to be divided into those groups which are to receive standard doses and those which are to receive test doses. It is important that randomisation is achieved by arbitrary selection and allocation to the different groups. Remember that we are dealing with quantal effects so it is not possible to give each group both standard and test doses.

**SAQ 1.5b**

In an insulin assay at least 96 mice are needed and are randomly distributed into 4 groups (two test and two standards).

It would be possible to arrange the four groups by catching the first 24 mice and designating them as 1, the next 24 as 2 and so on.

Can you give a fairly simple reason why this practice should *not* be adopted.

**SAQ 1.5b**

It is rather better to distribute the mice into the four groups sequentially, and to improve the randomisation even further not to allocate in the sequence 1,2,3,4,1,2,3,4 and so on but to use 1,2,3,4,4,3,2,1 etc. An even better practice might be to number each mouse and deal cards bearing the respective numbers into four piles to allocate the four groups. The tables of random numbers found in many sets of statistical tables, or random number computer programs, may also be used to allocate the individuals into groups. Whilst it has been stated that environmental conditions should be kept constant, this is not achievable in the absolute sense and further randomisation is often necessary in the location of animals during the test period. In the insulin assay the mice should be randomly located in the incubator to allow for any temperature variations, eg the bottom might be warmer than the top, there may be variations in accessibility of food and water and even differences in lighting might be significant.

## 1.6. EVALUATION OF ASSAYS

It might perhaps be obvious that the interpretation of results from these assays is going to be a complex matter because of the joint problems of small sample size and the generally poor precision (reproducibility) associated with bioassays and caused by the many variables of the procedures. Again some quite complex statistical methods can be involved in the analyses.

It is not the place of a discussion such as this to enter into these considerations in depth. However it should be noted that for most assays the data ought to fit certain criteria. For example a linear response (either directly or after mathematical manipulation of one or both of the parameters) is necessary. The observed responses to replicates of the treatments should be normally distributed about mean values, and the coefficient of variation (standard deviation expressed as a percentage of the mean) values obtained should be constant and independent of the dose levels. Further discussion and examples of the means of evaluation of assays is given in Part 2 of the text.

**Summary**

Bioassays depend on measuring the response obtained when a stimulus is applied to a biological system. The biological system may take a variety of forms which include the use of whole animals and animal tissues. Such bioassays are usually very sensitive and highly specific for the biologically active analyte contained in a sample. They are however expensive, precision is poor compared with alternative assay methods and there is considerable pressure from society to discontinue using animals in this way. To improve precision and give some degree of comparability between the results from different laboratories all bioassays must be carefully designed and every aspect of the methodology exactly specified.

**Objectives**

You should now be able to:

- describe the fundamental principles of bioassays;
- discuss the relative advantages of bioassays compared with physical and chemical methods;
- explain the particular problems associated with the use of animals or animal tissues in bioassays;
- explain what is meant by biological activity;
- describe the standards used in bioassays in general terms;
- describe the effects obtained in bioassays using animals and their tissues;
- discuss aspects of the experimental design of bioassays.

# 2. Microbiological Assays

### Overview

Basic concepts of microbiology which are pertinent to the operation of microbiological assays are introduced in this part of the text with the recommendation that further reading is undertaken if this subject is new to you.

The relative advantages of microbiological assays in terms of cost and analytical performance are compared with assays using animals and animal tissues before reference to actual methods are dealt with. The two principal approaches to microbiological assays are those using broth cultures (tube assays) and assays on semi-solid culture media (plate assays). Principles of these techniques and further aspects of experimental design are discussed.

## 2.1. INTRODUCTORY MICROBIOLOGY

### 2.1.1. Basic concepts and nomenclature

Microorganisms include the prokaryotic bacteria and blue-green algae (cyanobacteria) and the simple eukaryotic organisms, namely the protozoa, slime moulds and some fungi. The feature which all these organisms have in common is the ability to lead an independent existence as single cells.

A detailed knowledge of microbial taxonomy (classification, nomenclature and identification) is not necessary to perform microbiological assays. However, if microbiology is relatively new to you it is advisable to read through the introductory sections in one of the basic text books to acquire at least a working vocabulary of the subject. It is important that you understand the necessity of correct nomenclature when ordering stock cultures of microorganisms or when publishing information regarding microbiological assays. Just like higher animals and plants, microorganisms are identified by the binomial system of nomenclature which uses the genus and species names. In printed literature it is the convention for the names of organisms to appear in italics eg, *Staphylococcus aureus, Homo sapiens*. When the material is handwritten or italics are not possible on your particular typewriter the names should be underlined, eg Staphylococcus aureus. When a culture is ordered for assay purposes it is advisable to quote the name of the organism and a catalogue number. Names are changed from time to time and sometimes more than one strain of a particular organism may be used for the assay of the same compound eg several strains of *Lactobacillus leichmanii* have been used in vitamin $B_{12}$ assays. Many countries have one or more national type culture collections from which strains suitable for most published microbiological assays may be obtained for a small charge. For example in the United Kingdom test organisms may be obtained from the National Collection of Industrial Bacteria, Torry Research Station, Aberdeen, Scotland. Although yeasts, protozoa and algae can be used for microbiological assays most methods use bacteria which are generally easier to maintain. The methods described in the following text will tend to refer to bacterial cultures but many of the details are equally applicable to microorganisms in general.

It is also of importance that you have a basic knowledge of bacterial growth and metabolism, factors influencing these and the technical procedures involved in culturing bacteria since successful assays depend on them. Section 2.1 briefly reviews these topics, but once again you are advised to refer to basic microbiology texts if the subject area is new to you.

### 2.1.2. Bacterial growth

Bacteria multiply by an asexual process called binary fission which involves a limited amount of cell enlargement followed by division to give two daughter cells.

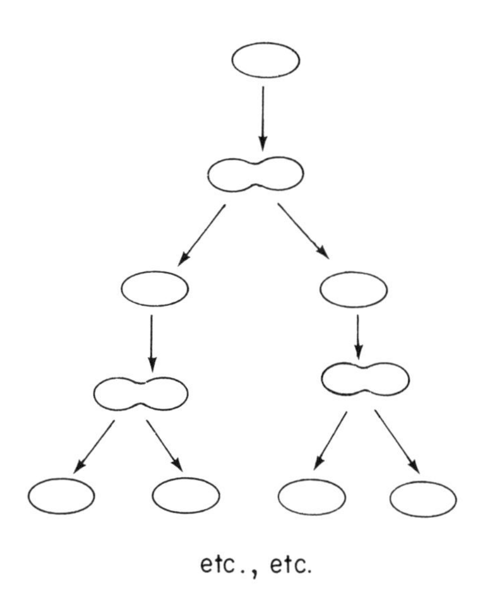


**Fig. 2.1a.** *Growth of bacteria by binary fission*

When there are equal periods between successive cell divisions a logarithmic or exponential rate of growth is seen.

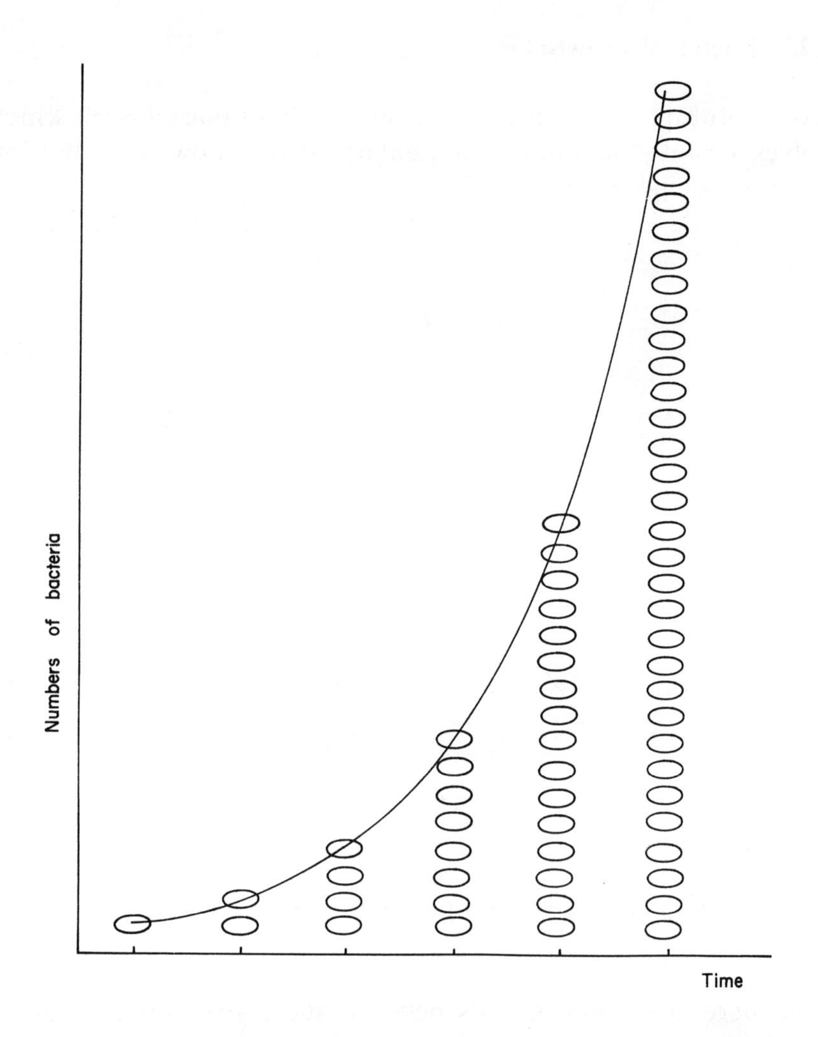


**Fig. 2.1b.** *Graph showing the logarithmic growth of bacteria*

During the logarithmic phase of growth of the culture, counts of the numbers of bacteria $b_1$ and $b_2$, taken at times $t_1$ and $t_2$ can be related by the formula:

$$b_2 = b_1 \times 2^n \qquad (2.1a)$$

where $n$ is the number of cell divisions which occurred in $(t_2 - t_1)$ minutes. Converting to a logarithmic expression we have:

$$\log b_2 = \log b_1 + n \log 2 \tag{2.1b}$$

which gives:

$$n = \frac{\log b_2 - \log b_1}{\log 2} \tag{2.1c}$$

Since log 2 is 0.3010, $n$ may be calculated by the simplified formula:

$$n = 3.322 \times (\log b_2 - \log b_1) \tag{2.1d}$$

The cell division took place over $(t_2 - t_1)$ minutes and therefore the generation time is equal to

$$(t_2 - t_1)/n \text{ (minutes)} \tag{2.1e}$$

**SAQ 2.1a**

The table shows numbers of *Escherichia coli* per $cm^3$ in a liquid culture medium over a period of incubation at 37 °C.

Convert the counts to logarithms to the base 10 and plot the logarithmic values against time on the graph.

Calculate the generation time for *Escherichia coli* over a period of incubation when there is a straight line relationship between the logarithm of numbers present and time.

| Time (min) | Count (no. per $cm^3$) | log (count) |
|---|---|---|
| 60 | $1.6 \times 10^2$ | |
| 120 | $2.3 \times 10^2$ | |
| 180 | $1.7 \times 10^3$ | |
| 240 | $1.0 \times 10^4$ | |
| 300 | $7.1 \times 10^4$ | |
| 360 | $4.7 \times 10^5$ | |
| 420 | $3.1 \times 10^6$ | |
| 480 | $2.0 \times 10^7$ | |
| 540 | $1.3 \times 10^8$ | |
| 600 | $2.5 \times 10^8$ | |

**SAQ 2.1a**

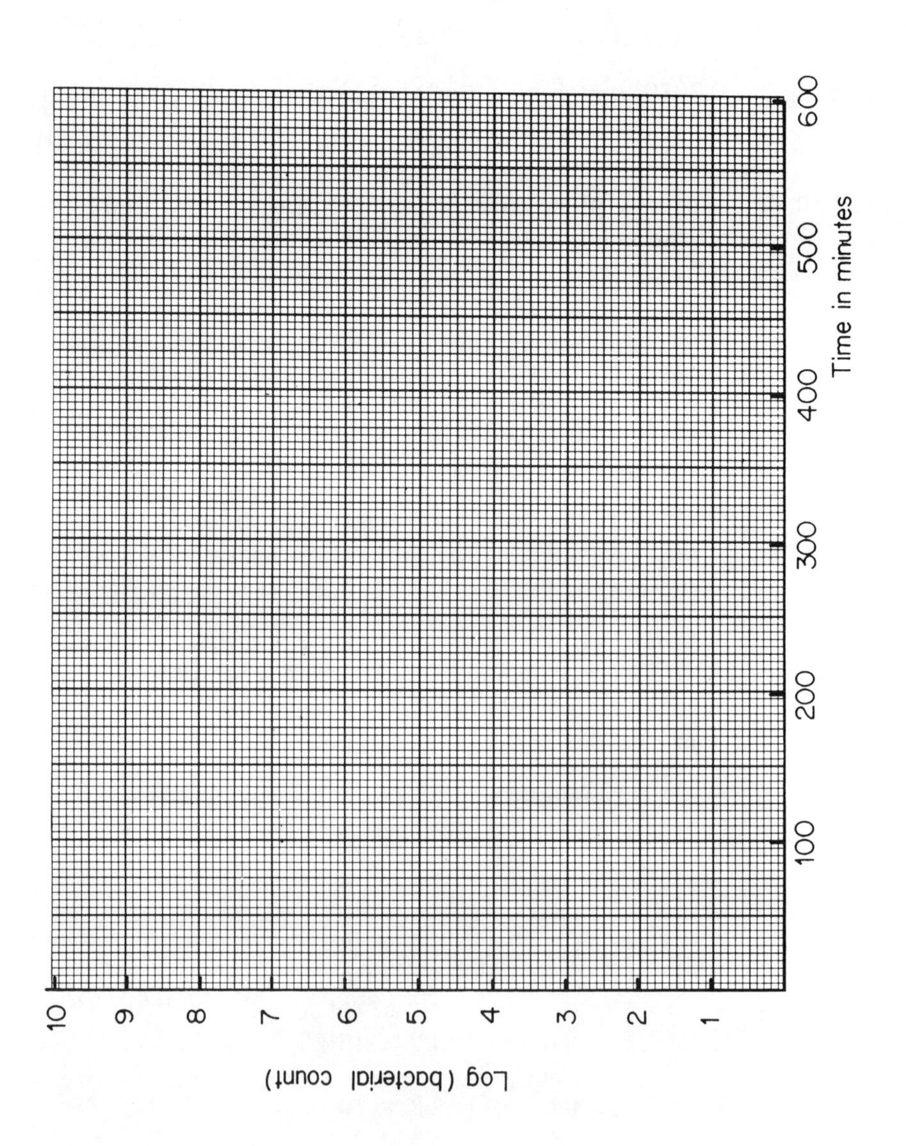
Time in minutes
100
200
300
400
500
600
Log ( bacterial count)
1
2
3
4
5
6
7
8
9
10

Though some species of bacteria have greater maximum growth rates than others, the form of the time course of growth is similar. After an initial lag phase while the bacteria adapt to the new environment, the population grows exponentially, ie the logarithm of the numbers increases linearly with time.

Laboratory methods are usually timed to function within the logarithmic phase of growth when the population is at its most uniform in terms of chemical composition of the cells, metabolic activity and physical characteristics. The growth rate decreases after several hours (ie growth is no longer logarithmic, Fig. 2.1c) and eventually the stationary phase is attained when no net change in the number of viable cells takes place. This is brought about by a variety of influences including the exhaustion of nutrients and the accumulation of toxic waste products. Eventually the stationary phase gives way to the phase of decline (death phase) which may take just a few hours in the case of a delicate organism or months to years with the more hardy species.

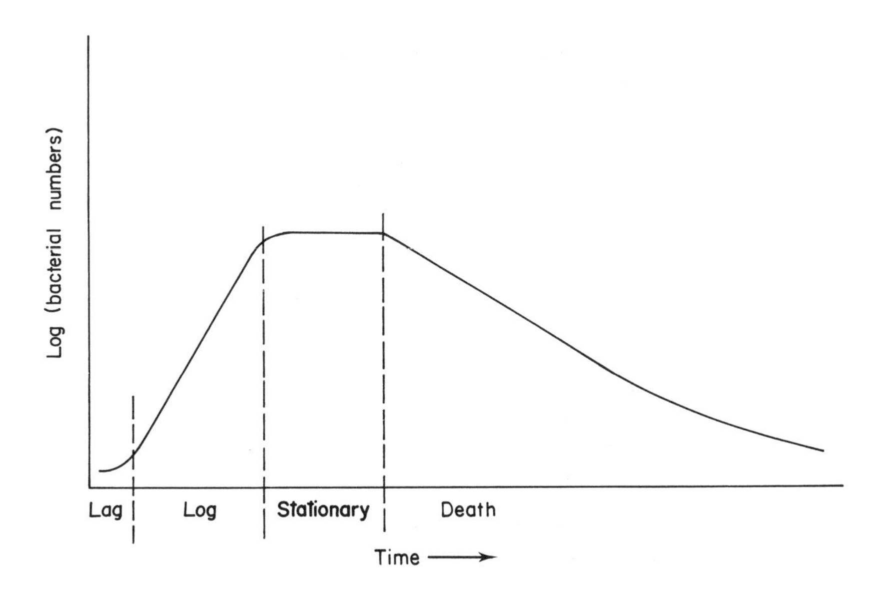


**Fig. 2.1c.** *A typical bacterial growth curve showing the four phases of growth*

### 2.1.3. Growth requirements

Autotrophic bacteria are able to grow in a culture which contains a simple selection of inorganic salts. This is in contrast to nutritionally exacting heterotrophic types that demand a complex mixture of organic substances.

In order to supply these nutrient requirements culture media range from defined synthetic media prepared from analytical grade chemicals to media which contain yeast or meat extracts and digests and have a complex, undefined compositions. In some cases culture media can be further enriched by the addition of body fluids such as blood or serum.

Culture media may be used in the liquid state (broth cultures) or in a fairly solid state produced by the incorporation of agar at a concentration of about 1 to 2% (w/v). Agar is a heteropolysaccharide derived from certain types of seaweed, it is not utilised as a source of nutrients by most bacteria thus providing an ideal gelling agent.

**SAQ 2.1b**

Mark with a cross whether each of the following is associated with culture in a broth medium or a surface-inoculated agar medium. You may assume that about the same number of bacteria are inoculated and that the volumes of broth and agar media are similar.

| | Broth medium | Agar medium |
|---|---|---|
| Greater availability of nutrients | | |
| Higher yield of bacteria | | |
| Growth of bacteria as colonies (clumps) | | |
| Greater sensitivity to the atmosphere | | |

In addition to the composition of the culture medium each species of bacterium requires the correct temperature, carbon dioxide and oxygen levels for optimum growth. A bacterium may be placed in one of three groups on the basis of its ability to grow over a certain temperature range. Psychrophiles grow at 0 °C or even less although they often grow more rapidly at temperatures of about 20 °C. Mesophiles grow more rapidly at temperatures in the range 25 to 40 °C. Most of the bacteria which are capable of causing infections in man (pathogenic bacteria) are mesophiles. The thermophiles grow well at higher temperatures from about 45 to 60 °C and some types have even been isolated from hot springs with temperatures around 100 °C.

The growth of many bacteria is encouraged at increased levels of carbon dioxide compared with atmospheric level of the gas. There are some species that will grow poorly or not at all unless the carbon dioxide tension is around 5 to 10% (v/v).

The oxygen requirements of bacteria and other living organisms vary considerably. It is possible to identify four groups of organisms on the basis of oxygen need:

- obligate aerobes that cannot grow unless oxygen is present;

- obligate anaerobes that will only grow if oxygen is absent from the environment. Some obligate anaerobes may die after even brief exposure to oxygen;

- facultative anaerobes which show optimum growth in the presence of oxygen, but can continue to grow in its absence. The fermentation process whereby ethanol is produced by yeasts is an example of a facultative anaerobe in culture;

- microaerophiles which grow best when the oxygen tension has been reduced below normal atmospheric levels.

Some bacterial species require a specific supply of nutrients or will only grow in a closely defined physical environment, eg pH range. Most however can tolerate a fairly wide range of conditions and these types are generally used in microbiological assays. Even these less-exacting species give variable growth rates and final yields unless the culture media and physical conditions for growth are carefully standardised and maintained throughout the assay procedure.

Optimum growth is usually taken to mean the most rapid growth rate over a relatively short period, but this may not result in the highest yield of cells. Fig. 2.1d shows that a *Bacillus* species gives the highest growth rate at 25 °C but a much greater yield of cells at 5 °C although almost six days are needed to achieve this.

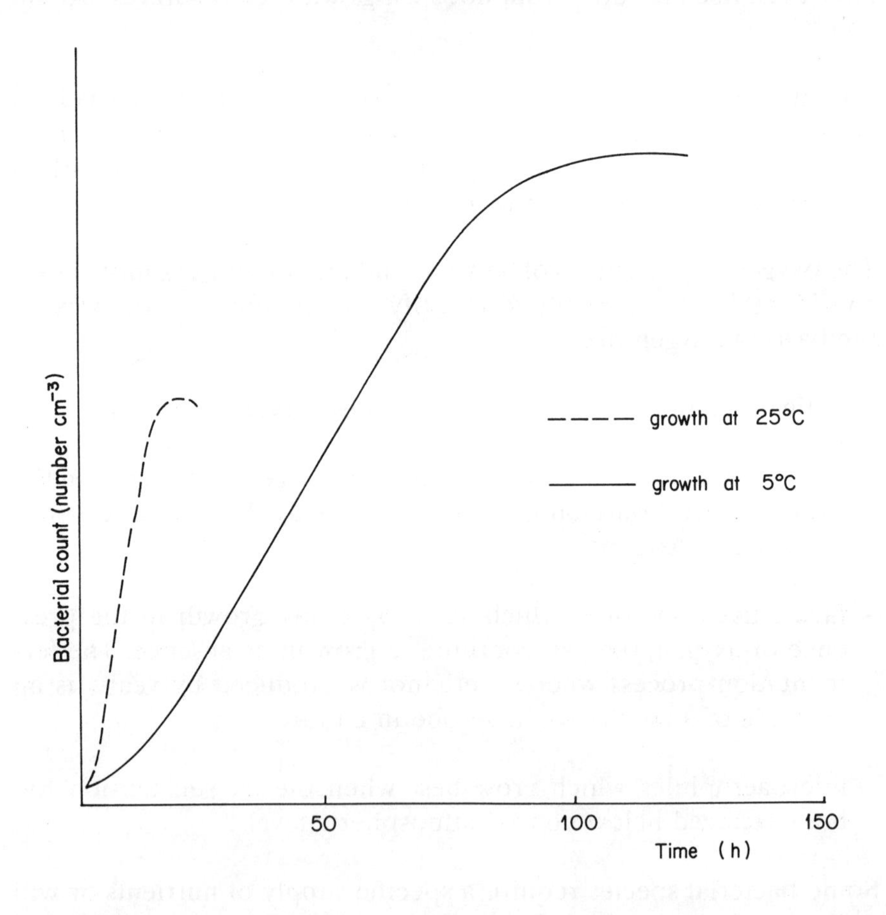


**Fig. 2.1d.** *Growth of a Bacillus species at 25 °C and at 5 °C against time*

Π From the information supplied in Fig. 2.1d would you classify the *Bacillus* species as a psychrophile, mesophile or thermophile?

The *Bacillus* species is a psychrophile since it has the ability to grow well at low temperatures in spite of the fact that it also grows at 25 °C.

Incidentally the use of a capital letter 'B' in the word '*Bacillus*' indicates that the organism is a member of the genus *Bacillus*. You will come across the word bacillus (small b) used to describe any bacterium that is rod-shaped irrespective of its genus.

### 2.1.4. Microbiological techniques and precautions

It is not appropriate to discuss sterilisation methods, aseptic techniques and the safe handling of microorganisms in this text. We strongly advise that you obtain some training and practical experience in a microbiology laboratory before setting up facilities for microbiological assays.

In many countries there are national bodies who determine the standards of design, specifications of equipment and working practices which must be achieved by microbiological laboratories. These bodies usually base their recommendations on a system of classifying organisms according to the degree of risk which they present to laboratory workers and the population at large. Although highly dangerous microorganisms are hardly likely to be used for microbiological assays, it is nevertheless important that you are aware of the risks, and can meet the approved safety requirements. In the United Kingdom publications by the Advisory Committee on Dangerous Pathogens provide guidelines on issues of safety and laboratory facilities.

## 2.2. COMPARISON OF MICROBIOLOGICAL ASSAYS WITH BIOASSAYS USING ANIMALS

Microbiological assays have several advantages over assays using animals. In Part 1 of the text we mentioned that the use of animals raises ethical problems and many countries, quite rightly, impose severe restrictions on such work. The provision, maintenance and staffing of facilities for experimental animals is also extremely expensive. Even inbred animals show variations in response to the same dose of analyte and it is necessary to perform all assays in replicate. The extent of replication is however limited by expense and a natural desire to keep the numbers of animals subjected to the assay to a minimum. Variation in assays using bacteria is not a problem since asexual reproduction leads to populations with minimal differences in the genetic composition of individuals. Occasional spontaneous mutations do occur, but the number of such mutant strains are usually insignificant.

Microbiological assays can be carried out without highly specialized and expensive equipment. The methods are therefore worth considering in situations where the numbers of tests are likely to be small and large capital expenditure on equipment etc cannot be justified.

One limitation of microbiological assays is that they can only be applied to analytes which either promote or inhibit microbial growth. Also no other substances must be present in the test samples which either promote or inhibit growth or modify the response to the analyte.

## 2.3. BASIC APPROACHES TO MICROBIOLOGICAL ASSAYS

The two most common approaches to microbiological assays are plate (agar diffusion) and tube (liquid media) methods. In both methods growth promoting or inhibiting potencies of test samples are compared with standard preparations of the analyte.

Evidence suggests that the choice of a plate or tube method for the assay of an analyte is more dependent on custom and practice plus the analyst's personal experience rather than on objective considerations of the advantages and disadvantages involved. For example, plate assays which require overnight incubation have been the preferred microbiological methods for the measurement of antibiotics in the United Kingdom and in many other countries over the past thirty years in spite of the availability of four hour tube methods that appear to be equally reliable.

Examples of practical procedures for plate and tube assays of penicillin are described in outline in the following two sub-sections. Although the two methods appear very different they are based on the following common principles:

- a comparison of a test sample of unknown potency with a standard preparation of the analyte;
- a measurement of the inhibitory effect of the antibiotic on the multiplication of the test organisms;
- a quantitative relationship between the response of the test sample and standard preparation.

### 2.3.1. Plate assay of penicillin

A small volume of a broth culture of a penicillin-sensitive strain of *Staphylococcus aureus* is mixed with sterile molten nutrient agar at 45 to 50 °C.

15 $cm^3$ of the mixture is poured into a sterile Petri dish and allowed to set.

6 mm diameter wells are cut in the agar using a cork borer or gel punch. Measured volumes of test samples and a series of dilutions of the penicillin standard are pipetted into different wells.

After a pre-incubation period at room temperature to allow diffusion of the antibiotic the plates are incubated overnight at 37 °C.

Growth of the organisms is evident from the turbidity of the medium whereas inhibition of growth by the penicillin is shown by clear areas around the wells.

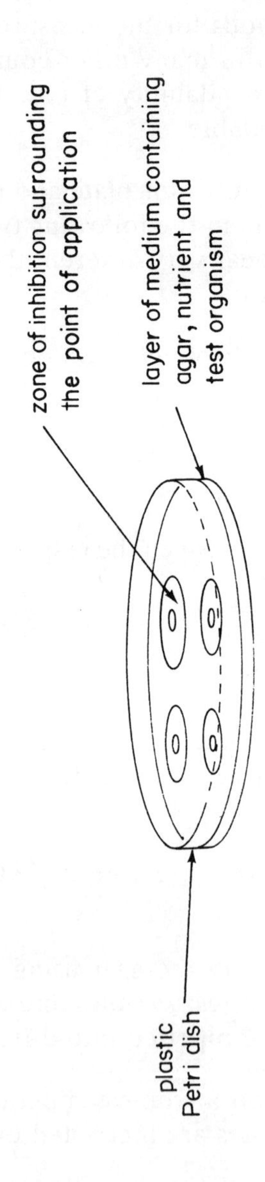


**Fig. 2.3a.** *Diagram of a plate assay showing inhibition zones around the wells*

The diameters of the zones of inhibition are measured and test zone diameters are compared with those of the penicillin standards.

### 2.3.2. Tube dilution assay of penicillin

Doubling dilutions (1/2, 1/4, 1/8, 1/16 etc) of a standard penicillin preparation are made in nutrient broth. It is important that the range of concentrations achieved by the dilutions includes the minimum inhibitory concentration (MIC) for the bacterium to be used in the test.

A similar set of doubling dilutions of the test sample is also made.

All the tubes are inoculated with a small volume of a broth culture of a penicillin-sensitive strain of *Staphylococcus aureus*.

The tubes are incubated at 37 °C overnight and then examined for turbidity which indicates the presence of growth (Fig. 2.3b). In this instance turbidity is recorded as present or absent by simple visual observation.

To provide an easier means of detecting growth by simple visual observation glucose and a pH indicator such as phenol red may be included in the nutrient broth. *Staphylococcus aureus* ferments glucose producing acid which lowers the pH and gives a colour change (red to yellow with phenol red) that is easily detected.

The basic premise used in the method is that the highest dilutions of both standard and test sample which inhibit growth must have the same potencies of antibiotic. The penicillin content of each standard tube can be found by multiplying the dose level of the stock solution by the dilution (expressed as a fraction).

Π Given that the stock penicillin solution used for the assay illustrated in Fig. 2.3b was 0.8 U $cm^{-3}$, calculate the penicillin concentration of the test sample.

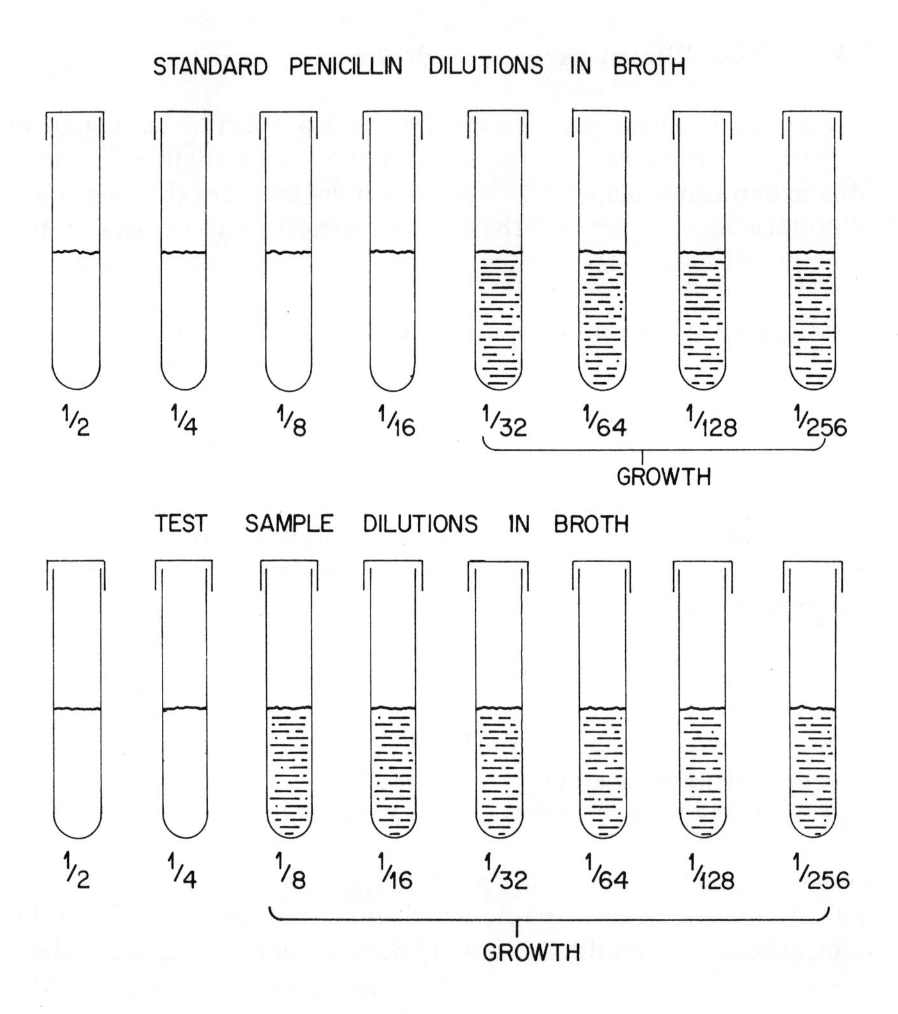


**Fig. 2.3b.** *Tube dilution assay of penicillin*

No growth occurs in the standard row up to the 1/16 dilution which contains $1/16 \times 0.8 = 0.05$ U cm$^{-3}$ of penicillin. The test sample showed no growth up to and including the 1/4 dilution. If we accept that the 1/16 standard dilution and the 1/4 test sample dilution have the same penicillin concentrations then the test sample must contain $0.05 \times 4/1 = 0.2$ U cm$^{-3}$.

Even if we accept for the moment that the conditions in standard and test rows are identical, the premise that the highest dilutions of both the standard and the test sample that show no growth contain the same potencies of penicillin does not stand up to close examination. For example in Fig. 2.3b it is correct to say that the MIC of penicillin for the *Staphylococcus* lies between 0.025 U cm$^{-3}$ and 0.05 U cm$^{-3}$ and between the concentrations represented by the 1/4 and 1/8 dilutions of the test sample.

It is, however, quite possible that the MIC was only just below 0.05 U cm$^{-3}$ (the 1/16 dilution of the standard) and only just above the activity represented by the 1/8 dilution of the test sample. If this was so then the result which you calculated for the test sample activity was almost half the actual value.

An approach that overcomes this problem is to take readings of all tubes using a spectrophotometer or nephelometer during the logarithmic phase of growth. These readings relate to the numbers of organisms in each tube and can be used to calculate the analyte activity over a continuous scale of values rather than in steps. This approach is described in greater detail in Section 2.5.

### 2.3.3. Assay of growth promoting substances

In plate and tube dilution assays of growth promoting substances (GPS) such as vitamins, the response is exactly the opposite to assays of growth inhibiting substances. Plate assays of GPS show zones of growth around the wells and in tube assays it is the lower range of dilutions of standard and test sample where growth occurs. Whether the assay of GPS is by plate or by tube dilution the following criteria must be satisfied:

- the test organism must be dependent for growth on the presence of the analyte;
- growth rate (or final yields) must be quantitatively related to the dose of the analyte in the medium;
- the analyte must be absent from the assay culture medium which otherwise should contain an abundance of the other nutrients which are required by the test organism;
- the culture medium used to grow the organism before the assay is performed should contain all the essential nutrients including adequate quantities of the analyte;
- the test organism must be washed before it is used in the assay to prevent any transfer of the analyte into the assay medium.

## 2.4. PLATE ASSAYS

### 2.4.1. Seeding the assay agar medium

Culture media used for plate assays contain 1 to 2% (w/v) agar according to the origin of the latter. At these agar concentrations media liquefies on heating to 100 °C and sets to a firm gel at about 45 °C. In addition to the agar, the medium must contain the nutrients required by the test organism, but remember that the exact composition also depends on whether the assay is of a growth inhibiting or growth promoting substance.

**SAQ 2.4a**

Is the composition of the assay medium the same or different from the culture medium which is normally used to grow the test organism in the assay of:

(*i*) growth inhibiting substances;

(*ii*) growth promoting substances?

The most common method of seeding (ie inoculating) the medium with bacteria is to mix a small volume of a suspension of the test organism with the molten assay medium at about 50 °C. The seeded medium is then poured into a Petri dish or a square assay plate and allowed to set at room temperature. This brief exposure to temperatures in the region of 40 to 50 °C should not affect the viability of the organisms drastically. However, since it is not possible to reproduce the method of seeding exactly with respect to the phase of growth of the organism, the numbers present and the exposure to high temperatures in the molten medium, zone diameters obtained from different batches of seeded agar medium should not be compared.

The seeded medium should be allowed to set on a level surface to obtain an even thickness of gel. Usually only one of the large 24 × 24 cm square dishes is required for an assay whereas several of the smaller (10 cm diameter) Petri dishes are generally needed. Care should therefore be exercised to ensure that each dish receives the same volume of medium to maintain uniform thickness.

**SAQ 2.4b**

In a plate assay, are zone diameters likely to increase or decrease as a result of:

(*i*) reducing the thickness of the medium;

(*ii*) increasing the number of organisms added to the assay medium?

**SAQ 2.4b**

An alternative method of seeding is to inoculate the actual surface of the solid medium. This may be achieved by spreading a small volume of a suspension of the test organism evenly over the surface with a glass rod or by pouring a suspension over the surface, decanting the excess fluid and drying the surface before applying the test and standard solutions. Both methods have been used with satisfactory results, but consensus opinion indicates a preference for the previously described pour plate method of seeding on the grounds that it is easier to standardise and gives clearer zones.

### 2.4.2. Application of standard and test samples to the media

Pipetting out the solutions presents no problems as long as the wells have been cut accurately. The use of a proprietary gel cutter is recommended but some practice is still needed to produce even wells with no splitting of the surrounding agar medium or lifting of the medium from the base of the plate.

The easiest type of gel cutter to use consists of a hollow metal tube of some 5 mm diameter fitted with a spring-loaded base section (Fig. 2.4a). The upper cylinder is attached to a vacuum pump. Having cut the agar medium the cutter is pressed down to seal the system and allow the plug of medium to be sucked out.

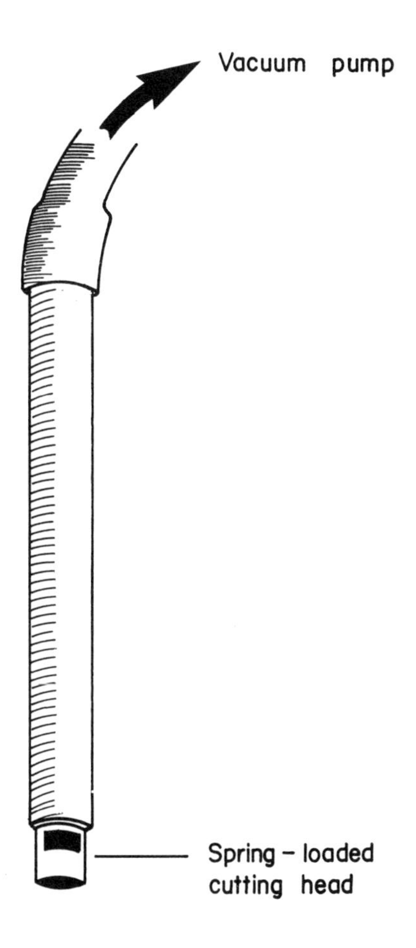


**Fig. 2.4a.** *Diagram of a commercially available gel cutter*

**SAQ 2.4c**

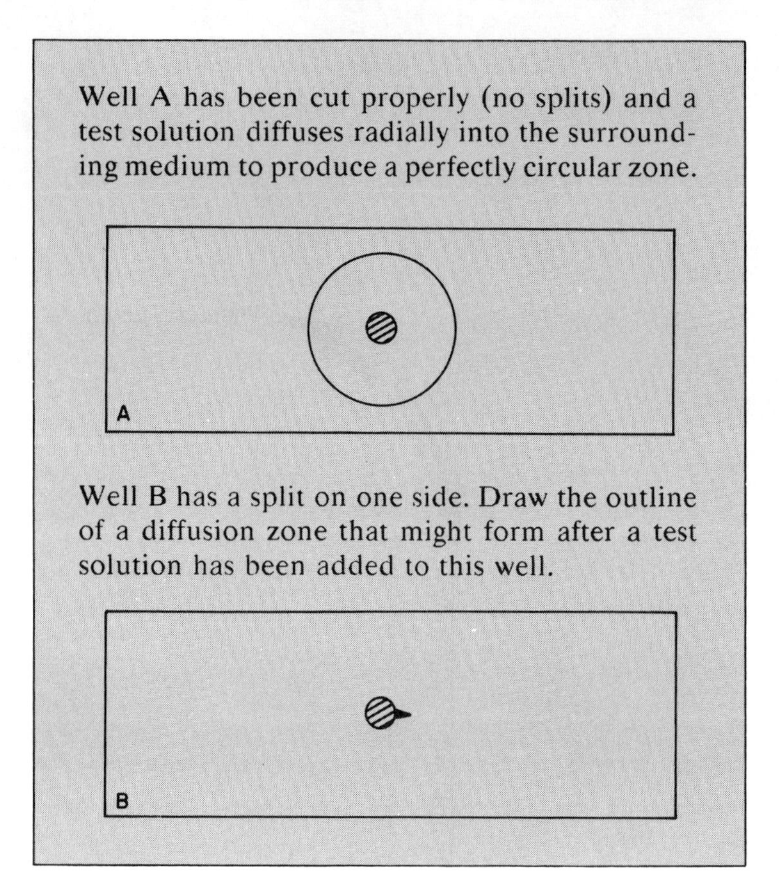


### 2.4.3. Measurement of zone diameters

Any obviously distorted zones should be discounted, but in practice few zones will have a perfectly circular periphery. To ensure that account is taken of small distortions in zones and to check the measuring technique it is advisable to take the mean of two zone diameters measured at right angles to each other (Fig. 2.4b).

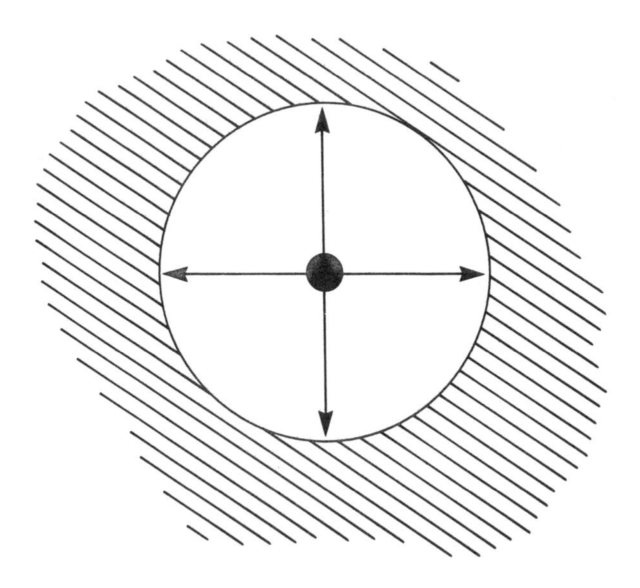

**Fig. 2.4b.** *Measurement of two zone diameters at right angles to each other*

The most popular alternative to wells as a means of applying standard and test solutions are small filter paper disks which are either dipped into the solutions and drained, or solution is pipetted onto them. The disks are then placed on the surface of the assay media.

### 2.4.4. Dose response curves

Different formulae are required to represent radial diffusion depending on whether the concentration of analyte in the reservoir (well, disk etc) remains constant or falls throughout the course of the assay.

**SAQ 2.4d**

The table shows zone diameters produced around wells filled with identical volumes of a series of gentamicin standards. Plot graphs of zone diameters versus gentamicin concentration (graph A) and zone diameters versus the logarithm (to the base 10) of gentamicin concentration (graph B)

| Zone diameter (mm) | Gentamicin ($\mu$g cm$^{-3}$) | log (gentamicin concentration) |
|---|---|---|
| 19.5 | 1 | |
| 20.8 | 2 | |
| 21.6 | 3 | |
| 22.2 | 4 | |
| 22.6 | 5 | |
| 23.0 | 6 | |

**SAQ 2.4d**

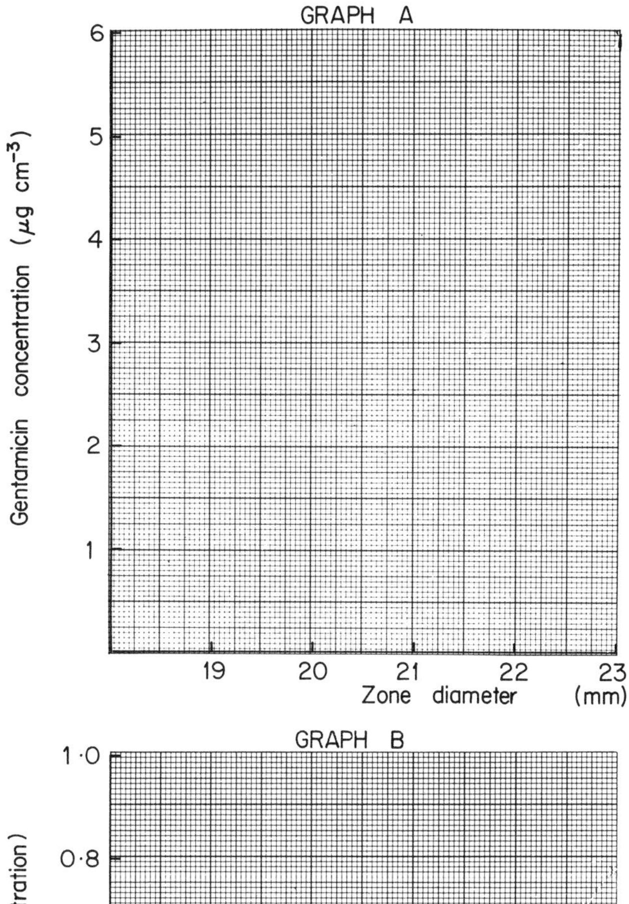
GRAPH A
6
5
4
3
2
1
Gentamicin concentration ($\mu$g cm$^{-3}$)
19
20
21
22
23
Zone diameter (mm)

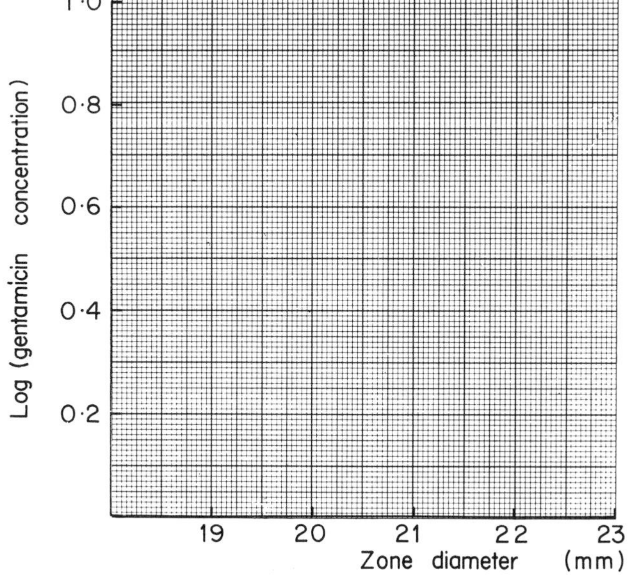
GRAPH B
1·0
0·8
0·6
0·4
0·2
Log (gentamicin concentration)
19
20
21
22
23
Zone diameter (mm)

In plate assays the logarithm of the analyte dose is directly proportional to the diameter of the zone (d in Fig. 2.4c) as long as the latter is large relative to the diameter of the well or disk (w). Doses which give small zone diameters should be avoided since well or disk sizes do become significant and log(dose) relates to the distance *r* rather than *d*.

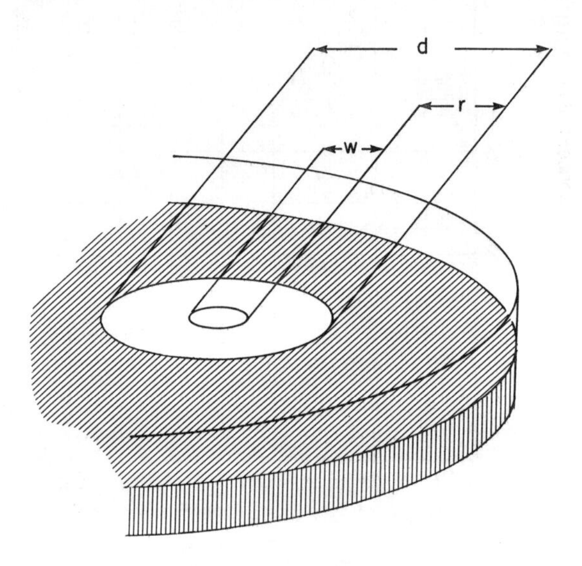


**Fig. 2.4c.** *Zone formation in a plate assay*

### 2.4.5. Other factors affecting zone diameters

The basis of plate assays is the relationship between the logarithm of the concentration of analyte and the zone diameter. You have already seen that the thickness of the agar medium and the size of the inoculum can influence zone diameters, but these are not the only factors. Factors which favour diffusion over growth will produce larger zones and vice versa. Since we wish to limit the only parameter affecting zone diameters to the activity of the analyte, it is important that all these factors be carefully controlled. Other factors which can affect zone diameters are as follows:

(*a*) Inherent sensitivity of the test organism.

The response of test organisms to growth-inhibiting or -promoting substances varies considerably from species to species and often from strain to strain of the same species.

(*b*) Physiological state of the test organism.

In some cases, eg members of the genus *Bacillus*, it is possible to use the test organism in either its vegetative or spore form. In the spore form active growth is delayed by the period required for germination. This effectively increases the pre-incubation period and gives rise to larger zone diameters than are obtained when vegetative cells are used.

Zone diameters are influenced by the phase of growth of vegetative cells at the time of inoculation into the assay medium. Smaller zone diameters tend to be produced by cells in the logarithmic phase. However, since the characteristics of an organism tend to be more predictable during the logarithmic phase it is generally this stage of growth which is preferred.

(*c*) Composition and condition of the assay medium.

Smaller zones are produced on a nutritionally rich medium in which rapid growth can occur. The moisture content of the medium also affects growth and therefore zone diameters. Drying out of gel media is usually more pronounced at the edges of plates and this can give rise to larger or mis-shapen zones.

(*d*) Prediffusion or preincubation periods.

It is important that all standards and test samples are applied to the assay medium over the shortest possible period to standardise the prediffusion times. This can be a problem if complicated assay designs are used which take a long time to set up. In such cases there may be an hour or more between applying the first samples to the system and the last, thus grossly affecting prediffusion times overall.

(*e*) Volume of analyte solutions

Zone diameters bear a direct relationship to the volume of sample which is applied. The method used to apply the standards and samples to the plates must be reproducible. Some workers apply measured (pipetted) volumes to wells while others prefer to fill each well exactly to the top. Generally when disks are used they are dipped into the solution and excess fluid drained off before applying to the surface of the seeded agar medium.

(*f*) Temperature of incubation.

Most rapid growth of an organism occurs at its optimum temperature and this leads to smaller zone sizes.

**SAQ 2.4e**

Can you explain why the optimum growth temperature is used for microbiological assays in spite of it being the temperature which gives smaller zone diameters and thus lower sensitivity. Reference to the graph showing how temperature affects the growth of a typical mesophilic bacterium may help with your explanation.

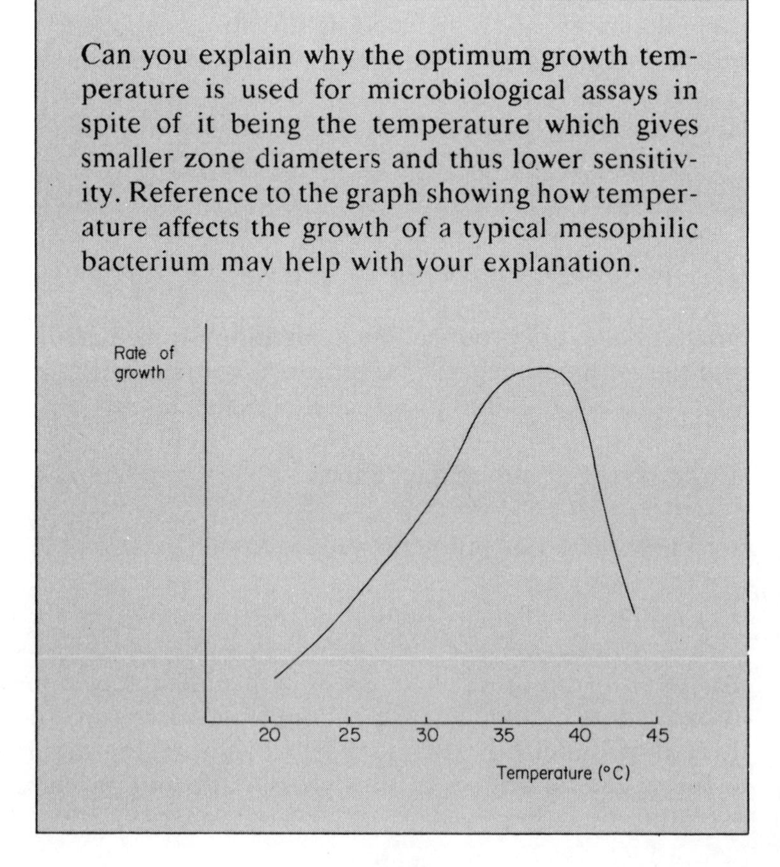

**SAQ 2.4e**

### 2.4.6. Assay design

A simple and perfectly legitimate approach to plate assays is to read the activities of test samples from a calibration curve. Fig. 2.4d shows a version of graph B from SAQ 2.4d redrawn on single cycle logarithmic-linear graph paper which is easier to work with than the type of paper which you used. The gentamicin activity of a test sample giving a zone diameter of 21.2 mm is easily read off the graph as 2.4 $\mu g\ cm^{-3}$.

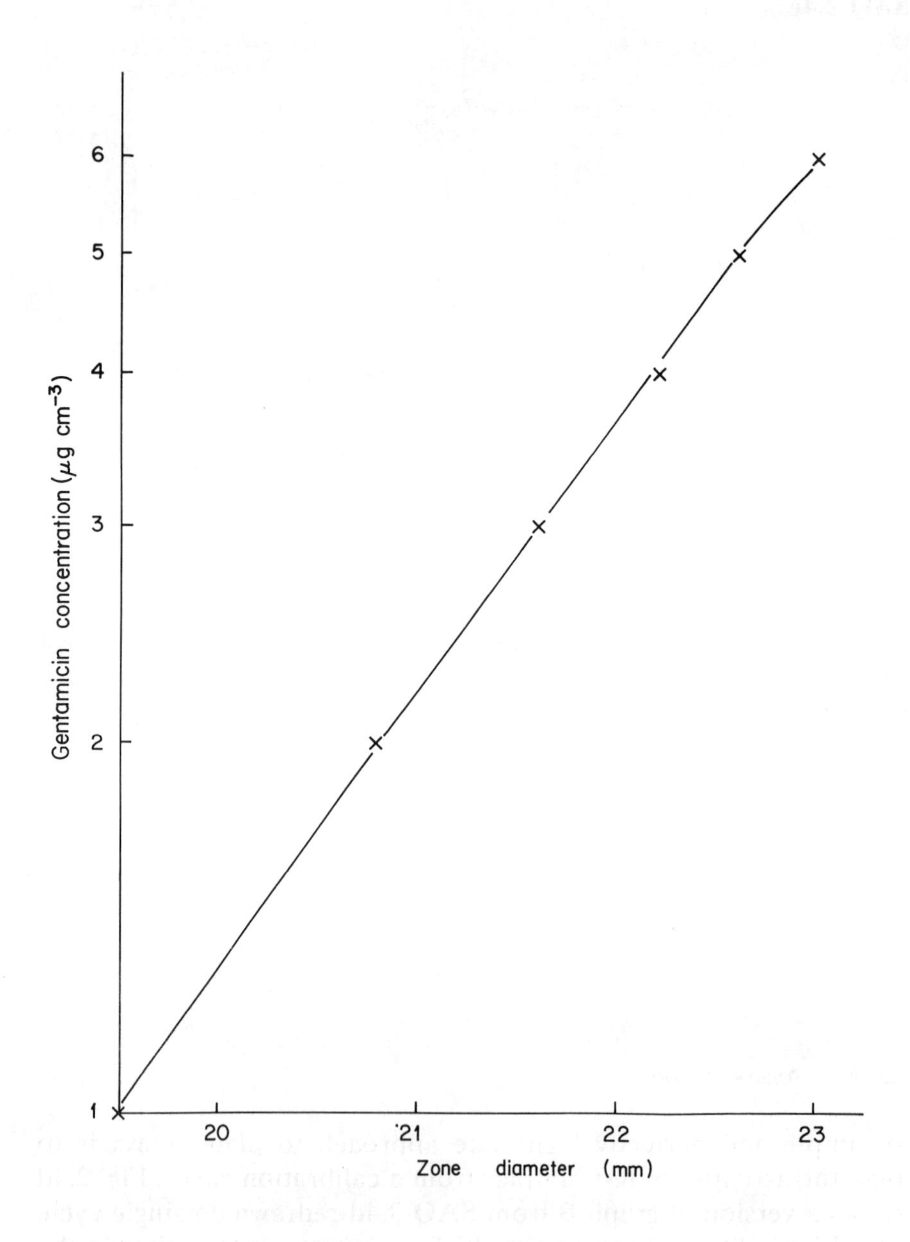


**Fig. 2.4d.** *Calibration curve for the plate assay of gentamicin. Note logarithmic scale of vertical axis*

The preparation of a calibration curve each time that an assay is performed necessitates the use of several standards. In practice this places a practical limitation on the degree of replication which is possible. Alternative designs do exist which give better results for the same practical effort and provide an indication of the validity of the assay. Some of these designs are discussed later.

### 2.4.7. Balanced assays

Balanced assays are designed to include the same number of tests and standards at equivalent dilutions. They are based on the characteristics of the log(dose) response rectilinear relationship which means that a plot of log (test sample) at two or more specified dilutions will be parallel to the plot of standards at the same dilutions. A minimum of two test and standard doses are required to plot response lines and it is this number which is used in the 2 + 2 assay design.

The 2 + 2 assay is set up using as many replicate Petri dishes of assay medium as is possible, with each dish containing two standard and two test solutions at the same specified dose ratios. The dose ratio refers to the relative concentration of the two doses. For example, if both the test and the standard are used undiluted and at a 1/4 dilution the dose ratio is 4 : 1. The assay is set up in exactly the same manner as previously described and zone diameters are measured. The log(dose) versus the mean diameters is plotted for the standard and a plot is also made of the two test samples as if they had the same analyte activity as the high and low standard doses.

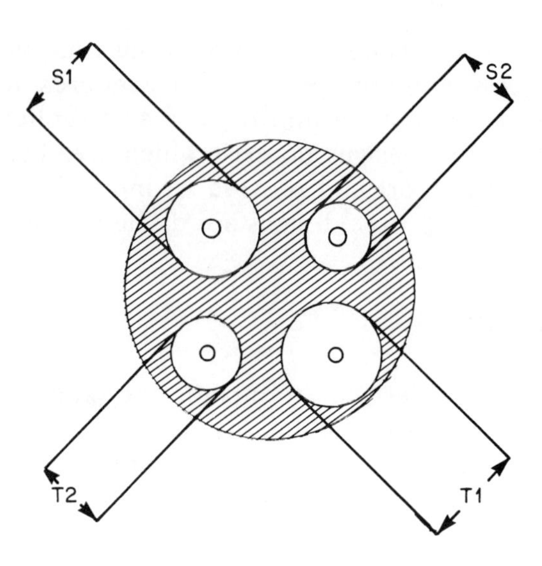


**Fig. 2.4e.** *A 2 + 2 assay design. S1 = diameter of high standard dose, S2 = diameter low standard dose, T1 = diameter of high test dose, T2 = diameter of low test sample dose*

**SAQ 2.4f**

The results of a 2 + 2 assay with *Bacillus subtilis* as the test organism and streptomycin at a ratio of 4 : 1 are tabulated.

| Treatment | Mean zone diameter (mm) |
|---|---|
| High standard ($80\ U\ cm^{-3}$) | 20.40 |
| Low standard ($20\ U\ cm^{-3}$) | 16.05 |
| High test sample used undiluted | 20.93 |
| Low test sample at 1/4 dilution | 16.50 |

⟶

**SAQ 2.4f (cont.)**

Plot log (standard dose) against the zone diameter and draw a straight line between the two points. Plot the test sample values as if the undiluted preparation contained 80 U $cm^{-3}$ and the 1/4 dilution 20 U $cm^{-3}$ against their respective zone diameters (20.93 and 16.50 mm)

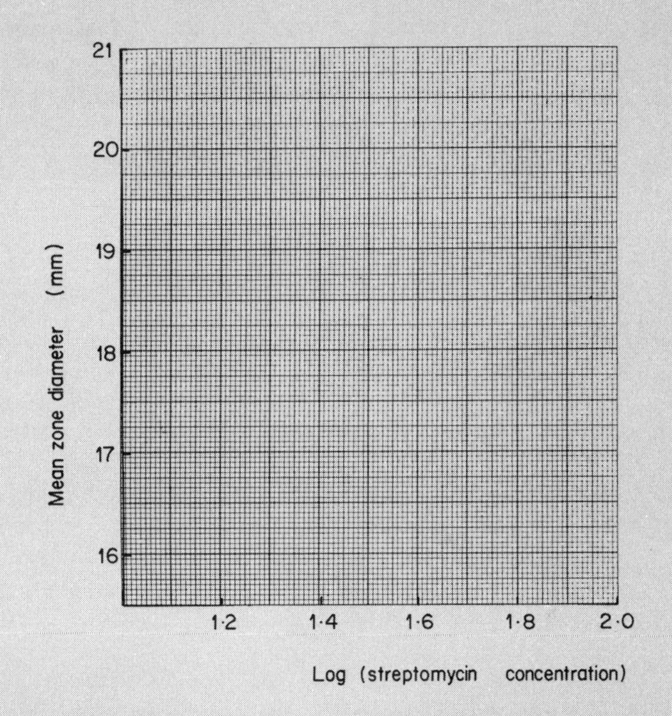


Note that your lines should be almost parallel which indicates that the assay was valid. Lack of parallelism is associated with invalid assays, and may for example indicate that the organism is responding differently to the test and sample materials. Measure the average distance ($M$) from the test to the standard line in terms of (streptomycin dose). This value is the logarithm of the relative dose of the test to the standard. Its antilog is called the potency ratio. ⟶

**SAQ 2.4f (cont.)**

$$\text{potency ratio} = \frac{\text{test dose}}{\text{standard dose}} = \text{antilog } M$$

The product of the potency ratio and the high standard dose gives the dose values of the undiluted test sample:

Test dose = potency ratio × high standard dose.

Calculate the potency ratio and the streptomycin level of the test sample.

The use of two test and standard doses shows parallelism but does not indicate whether the log(dose) response is linear. Both parallelism and linearity are checked in balanced assays where three or more (3 + 3, 4 + 4 etc) doses are used. The reliability of the assay also increases with the number of doses. However, with the larger assay designs, as with the use of calibration curves, the workload is markedly increased. In practice larger assays are usually confined to:

- new methods, when there is a need to establish linearity;
- methods where substances may be present in the test samples other than the analyte which could affect the result, eg other inhibitors of the test organism.

**SAQ 2.4g**

An analyst recognised the validity of calculating results using mean values derived from a large number of replicate tests. She set out ten replicate Petri dishes for 2 + 2 assays on each of ten different test samples making one hundred dishes in all. Her sequence of pipetting out the solutions into wells was in batches as follows:

*First*. All the high standard doses, ie 10 repeated manipulations from the same bottle.

*Second*. All the low standard doses, ie 10 repeated manipulations from the same bottle.

*Third*. The high test sample doses, ie 10 repeated manipulations from each of 10 bottles (100 in all).

*Fourth*. The low test sample doses, ie 10 repeated manipulations from each of 10 bottles (100 in all) ⟶

**SAQ 2.4g (cont.)**

On completion of the work all one hundred dishes were placed in an incubator without further delay and zone diameters were measured after 24 hours. Mean zone diameters were used to plot out the graph.

The analyst stated that this method of working was less likely to result in errors, such as placing the wrong solution in a well, and it also enabled the process to be completed in the shortest time. These two reasons seem perfectly acceptable, but there is a basic flaw in her approach which we would like you to identify. You may wish to consider that each pipetting sequence from a stock bottle to a well on the dish takes an average of 12 seconds and additional time must be added on for changing from one bottle to another, changing pipette tips etc.

### 2.4.8. Use of large plates

The larger assay designs (3 + 3, 4 + 4 etc) can be carried out on small Petri dishes but a single set of treatments requires several dishes, since one Petri dish normally takes only 4 or 5 wells. Treatments have to be randomly distributed in the different dishes and there is the possibility of variation in medium thickness between different dishes. On the whole, it is preferable to perform larger assays in large plates.

Large plates can accommodate many more wells which increase the number of doses and allows greater randomisation. The wells are normally arranged symmetrically in rows and columns, eg 8 rows and 8 columns giving a total of 64 wells.

In a single large plate it is easier to ensure that the medium is of uniform thickness but there are still problems, in particular:

- the large number of wells make a significant difference between the times at which first and last wells are filled so that prediffusion from those wells which are filled earlier gives them disproportionately larger zones; (this aspect of design was exemplified in SAQ 2.4g);
- there is a tendency for the medium at the edges of large plates to dry out with the result that zones around the wells at the edges of the plate may be large compared with zones towards the centre.

These variables can be overcome by filling the wells both:

(*a*) systematically *by position*, eg from top left, row by row, to end at bottom right (Fig. 2.4f).

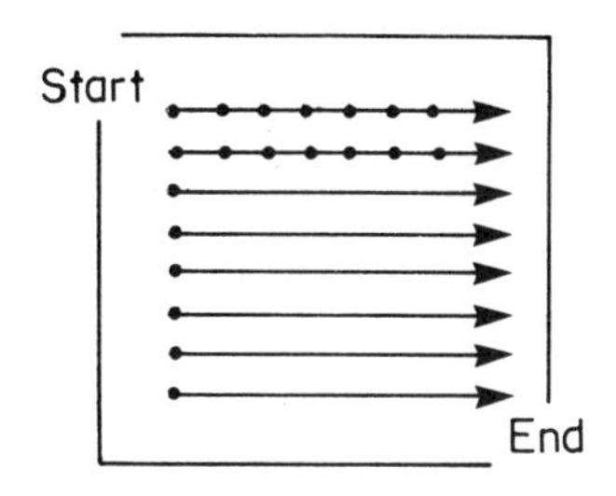


**Fig. 2.4f.** *Systematic filling of the wells on a large plate by position*

Other sequences are equally good as long as they are systematic by position, eg bottom left, column by column, to end top right;

(*b*) at equal time intervals filling the well in this way introduces a systematic bias which can be completely compensated by allotting treatments, to wells according to a Latin square design.

### 2.4.9. Latin square design

The concept of the Latin square design was introduced in Section 1.5. It is a block design restricted in that each treatment must appear once in each column and once in each row, but otherwise the sequences are completely randomised. Randomisation helps the laboratory worker avoid the bias caused by a subconscious desire to make replicates agree. The 8 × 8 Latin square design shown in Fig. 2.4g might be suitable for a large plate microbiological assay but if the laboratory is regularly performing this test it is quite possible for staff to remember the pattern of treatments. The influence of subconscious bias may be reduced by the use of a set of cards each of which displays a different 8 × 8 Latin square design. A laboratory worker selects one card at random from the set and then performs the assay using the design shown. By the use of this simple expedient the chance of subconscious bias affecting the results is extremely remote.

The 8 × 8 design enables the analysis of 8 'treatments' and it may be employed in a variety of ways according to the precision required, the number of samples for analysis and the extent to which it is thought necessary to test the validity of the analysis; thus, in order of decreasing precision:

(*a*) 1 test at 4 dose levels
+ 1 standard at 4 dose levels (ie a 4 + 4 assay)

(*b*) 2 tests at 2 dose levels each
+ 2 standards at 2 dose levels each

(*c*) 3 tests at 2 dose levels each
+ 1 standard at 2 dose levels

giving a total of 8 treatments in each case.

Latin squares of 9 × 9, 12 × 12 etc., are also used. Smaller squares are best avoided since the analysis of variance usually applied assumes each error to be normally distributed; an assumption less likely to be justified in smaller designs.

'Quasi-Latin' squares are frequently employed when it is necessary to analyse so many samples that the number of treatments is greater than the number of rows and columns. For example, an 8 × 8 square for 16 treatments in which each treatment appears once in every 2 rows and 2 columns, would enable the assay of 6 tests and 2 standards each at 2 different dose levels, It should be appreciated that opting for the quasi-Latin square must be at the expense of precision.

In microbiological plate assays, the Latin square design enables a statistical analysis of, and allowance for, variation over the test area, divided into rows and columns. The Latin square has many applications other than for bioassays. It is not just restricted to spatial situations but can be applied to any experiments with 3 sets of variables eg technicians, instruments and samples for analysis in quality assessment investigations.

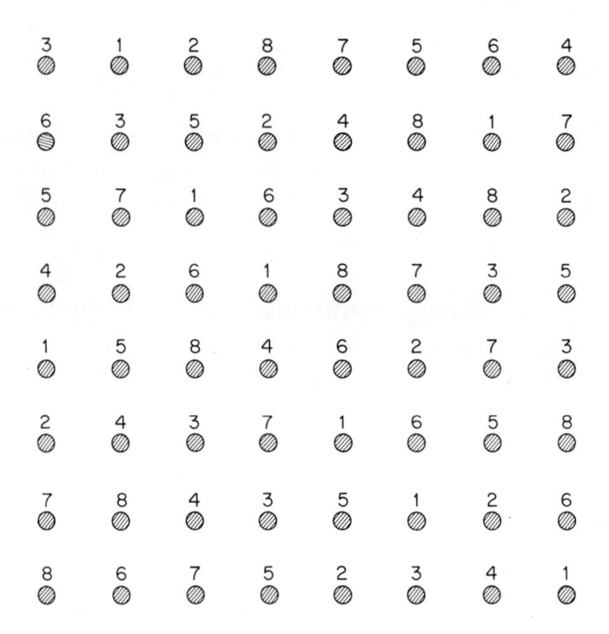


**Fig. 2.4g.** *Typical example of an 8 × 8 Latin square design*

Figure 2.4g shows a typical example of an 8 × 8 Latin square design which might be applied to a large microbiological assay plate containing a total of 64 wells. For a 4 + 4 assay, for instance comparing one test example with a standard, each at 4 dose levels a total of 8 treatments are involved. The 8 treatments (4 test and 4 standard doses) are numbered arbitrarily 1 to 8 and the wells are filled according to the pattern shown in Fig. 2.4g. It is essential that wells are filled in order of their position on the plate and not all the number 1 treatments followed by all the number 2 treatments and so on. The latter approach was that adopted by the analyst in SAQ 2.4g with unfortunate consequences arising from disproportionate prediffusion times.

## 2.5. TUBE ASSAYS

The tube assay of penicillin which was described in Section 2.3.2 does not provide a continuous scale of measurement and must be regarded as semi-quantitative at best. A truly quantitative response is obtained in tube assays where the amount of growth in the standard and test sample tubes are compared. The comparison can be made by measuring turbidity using a spectrophotometer or by measuring light scattering with a nephelometer in which the photodetector is positioned at an angle (usually 90°) to the incident light source.

As with plate assays it is important to maintain uniform conditions to ensure that the density of growth relates to the activity of the analyte in each tube. Factors such as the size of the inoculum, the growth phase of the test organism, media composition apply equally to tube assays as to plate assays, but with the former, incubation conditions may be more critical. It is extremely important for the validity of the assay to ensure that conditions are absolutely identical in each of the tubes. The following techniques have been used to obtain more uniform incubation condition in tube assays:

- broth media are chilled so that growth does not start until the tubes are transferred to the water bath;

- physically identical tubes (immersed in the water to the same depth) are used to ensure that heat transfer conditions are the same;

- large capacity, stirred water baths are used so that all the tubes are incubated at the same temperature regardless of their position;

- growth is terminated by simultaneous immersion of all the tubes in an 80 °C water bath to ensure the same incubation time (this is especially important with the short assay methods, eg 4 hours).

The vitamin $B_{12}$ method devised by Brian Kelleher and his colleagues at St. James's Hospital in Dublin is an example of a modern quantitative tube assay. They state a preference for a microbiological assay of the vitamin rather than the more popular radioisotope dilution methods since the latter may give rise to clinically misleading results.

The test organism used in the assay was developed from *Lactobacillus leichmanii*, (NCIB 8117) obtained from Torry Research Station, Aberdeen. The bacterium has a natural resistance to the antibiotic colistin sulphate, and this resistance was strengthened by growing the organism in broths containing increasing concentrations of the antibiotic. The organism is stored in a broth containing 0.5g $dm^{-3}$, and the assay is performed using medium containing 0.1g $dm^{-3}$ colistin sulphate. This decrease in antibiotic level causes the organism to grow rapidly, ensuring a relatively short incubation time, while still maintaining a sterile environment. The design of the assay obviates the need for stringent aseptic working practices – one of the principal disadvantages of microbiological assays.

The test organism is grown for 16 hours at 37 °C in a broth which contains colistin sulphate and all the necessary nutrients including vitamin $B_{12}$. The broth is mixed 50/50 with 80% glycerol and 1 $cm^3$ aliquots are stored at −70 °C. One vial of frozen organism is thawed for each assay, 100 $\mu$l of the inoculum is added to each litre of vitamin $B_{12}$ assay medium, and after thorough mixing, the inoculated medium is dispensed into tubes containing extracts of the test sera. A range of vitamin $B_{12}$ standards is included in each assay, and as a check on the accuracy of the assay, several sera of known $B_{12}$ levels are also included in each assay. The assay tubes are incubated in a 37 °C incubator for 42 hours.

Absorbance readings are taken in a spectrophotometer at a wavelength of 595 nm against a blank of uninoculated medium. A calibration curve of absorbance against concentrations of vitamin $B_{12}$ is drawn (Fig. 2.5a), the vitamin $B_{12}$ values of the known sera are checked, and finally values are calculated for the unknown sera. Since the response of the organism is not linear it is important to prepare a calibration curve for each assay. Using this assay vitamin $B_{12}$ values in the range 0–1000 pg $cm^{-3}$ serum can accurately be determined.

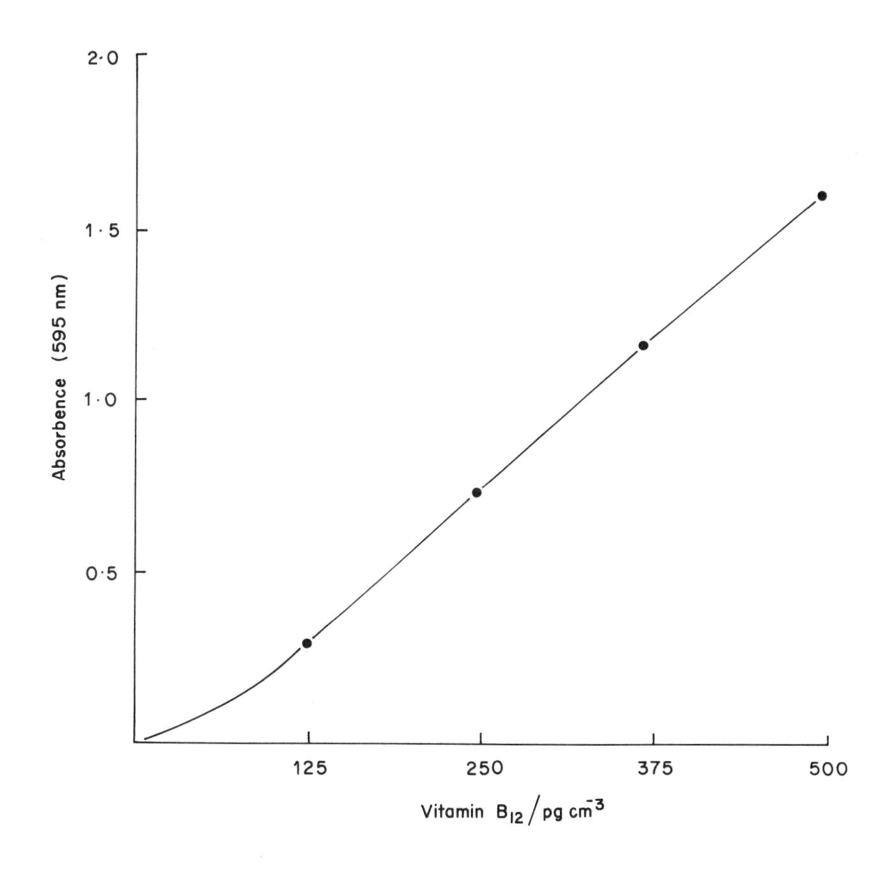


**Fig. 2.5a.** *Calibration curve of absorbance against concentration of vitamin $B_{12}$*

The convenience of frozen inocula, which are used once and then discarded, coupled with the use of an antibiotic-resistant organism reduces the microbiological expertise required to a minimum.

**Summary**

Microbiological assays depend on the ability of an analyte either to inhibit or to promote the growth of a selected microorganism. It is important that growth conditions should be carefully standardised to ensure that the analyte is the only factor which influences growth. A better awareness of the optimum conditions can be achieved by understanding the basic principles of microbiological techniques.

Assays may be performed in liquid (broth) culture media by comparing growth of the microorganism in the presence of standard preparations with the growth that occurs with test samples. An alternative to broth culture methods is the use of an agar gel media where the analyte relates to zones of inhibition or growth around the wells or disks according to whether the analyte is a growth inhibiting or promoting substance.

**Objectives**

You should now be able to:

- describe the characteristics of bacterial growth in broth media;
- calculate the generation time during the logarithmic phase of growth;
- describe the factors which can affect growth rates;
- discuss the relative advantages of microbiological assays compared with those of assays which use animals or their tissues;
- explain the criteria required for the assay of growth inhibiting and growth promoting substances;
- describe the principles of tube and plate assay techniques.

# 3. Introductory Immunology

## Overview

This Part of the Bioassay Unit will consider those aspects of basic immunology which are necessary to understand the use of antibodies in the measurement of biological materials.

## Introduction

In this Part we will cover some of the basic concepts of immunology necessary in order to understand the techniques discussed in the remainder of the Unit. Antibody production is part of the immune defence system of the body, the study of which, (immunology), is a very important and rapidly developing subject. Unfortunately it is also a very complex one. The immune system is not the only defence available to the body and antibody production is not the only function of the immune system. Here, however, only such immunology is included as is necessary to an understanding of this Unit. If you are going to become deeply involved in immunological methods, you may need a wider reading in immunology and the Bibliography lists texts suitable for further study.

## 3.1. ANTIBODIES AND ANTIGENS

The introduction of a foreign substance into the body may induce the formation of a special protein called an antibody. Antibodies circulate in the blood and can bind with the particular substances which invoked their production in the first place. This binding occurs with a high degree of selectivity, ie an antibody will bind to that substance but none, or at least very few, others and it is this selectivity in the binding action of antibody molecules which makes them so useful as analytical reagents. Moreover it is possible to obtain an immense range of different antibodies each of which will, in theory at least, bind just one substance. The product of the binding of antigen to antibody is called an immune complex and the fact that this can take the form of a visible precipitate is of use in some analytical techniques, although for the majority of immunologically-based techniques with a value in chemical analysis, it is simply the act of binding that is of value as will become evident later.

The materials which can bring about antibody production are called antigens, and different antigens give rise to different antibodies. The antibody corresponding to a particular antigen appears in the blood only after exposure to that antigen either naturally in the course of infections etc, or artificially, as for example following the experimental or prophylactic inoculation of an animal or human being.

### 3.1.1. The natural role of antibodies

Antibodies help protect against infectious diseases, that is those caused by invading microorganisms such as bacteria and viruses. Surface structures and/or products of these invading microbes act as antigens and stimulate cells called lymphocytes to produce the antibodies. This is the so-called adaptive immune response. These antibodies bind to the microbial antigens in a way that neutralises the harmful effects of the microbial products and facilitates the destruction of the microbes themselves. Antibodies can thus provide some defence against a disease.

### 3.1.2. Response to non-microbial antigens

Although the purpose of antibodies is to protect against disease-producing microbes, antibodies can be produced experimentally by the injection of many substances which are quite unconnected with these organisms. We can take advantage of this to produce antibody to our own requirement. Simply by injecting (immunising) an animal with a substance of interest, providing it is antigenic it should be possible to produce (raise) antibody which will bind that substance and which can be isolated from the animal for a variety of uses, eg targetting of drugs, neutralization of toxins, diagnostic assays, and the removal of cells by affinity chromatography.

### 3.1.3. The second response to the same antigen

In the case of many infectious diseases, those individuals who have had the disease once and survived prove much more resistant to the same disease if they are exposed to it again. This is, in part, the result of an ability to produce rapidly a large amount of the appropriate antibodies on a second exposure to the same set of microbial antigens. This time the invading microorganism is more effectively neutralised and less, or none, of the disease symptoms develop. The individual is said to have immunity.

This augmented immune response on a repeat encounter with the same antigen(s) is the basis of artificial immunisation (vaccination) against an infectious disease. Here the idea is to obtain the immunity resulting from purposeful exposure to the antigen but without being subject to the symptoms of the disease. For this immunisation the microorganism or its products may be modified in such a way that they retain their antigenicity but lose the ability to produce disease. When injected, this preparation (vaccine) generates an immunity to subsequent infection but does not produce the symptoms of the disease.

Once someone has acquired an immunity against a disease either by having had the disease or having been immunised against it, the effect lasts a long time, perhaps for as long as 10 years. Likewise an augmented production of experimental antibody can be obtained even when the second dose of an experimental antigen is given a long time after the first. Another property therefore of the immune system is its long 'memory' of those antigens which it has already encountered and this together with the increased yield following repeated injection is of great value in the production of antibodies for experimental purposes.

To summarise then an individual or experimental animal will have a number of antibodies in the blood circulation due to recent contact with foreign antigenic materials, and will have the potential of rapidly producing others using the cellular memory of previous encounters. In addition the individual should be able to produce reasonable quantities of entirely new antibodies should the body be challenged by a new antigenic material.

**SAQ 3.1a**

Define the terms

(*i*) antigen;
(*ii*) antibody;
(*iii*) response;
(*iv*) immune memory.

**SAQ 3.1a**

**SAQ 3.1b**

Complete the following sentences with the numbers of words indicated by the lines.

(*i*) When antibody and antigen bind, they form an ________ ________.

(*ii*) Antigens and antibodies combine with a high degree of ________________.

(*iii*) The purpose of antibodies is to help protect against ________________ diseases.

(*iv*) The quantity of antibody produced following a second exposure to an antigen is ________________ compared with that produced on first contact.

(*v*) Antibody can, under suitable conditions, be produced to antigenic substances of experimental and commercial interest by injecting them into an animal, a process called ________________.

**SAQ 3.1c**

What do you think the term anti-insulin antibody means?

**SAQ 3.1d**

Answer *true* or *false* to each of the following:

(*i*) the potential to produce a great diversity of antibodies is inherent in the immune system; T / F

(*ii*) the production of a particular antibody is the result of exposure to an antigen; T / F

(*iii*) the diversity of antibodies which can be produced benefits the animal or person because antibodies can be produced against a wide range of bacteria and viruses, and therefore helps resist infectious disease; T / F

(*iv*) the potential to produce antibodies of a particular specificity is the result of exposure to an antigen. T / F

## 3.2. CHARACTERISTICS OF ANTIGENS AND ANTIBODIES

When the term antigen was introduced, immune responses to infectious diseases were thought to result solely in antibody production. Later, when it was realised that additional, complex, cellular (cell-mediated) responses were also possible, a new term 'immunogen' was introduced, this term implying any substance capable of inducing any type of immune response, not just the production of antibodies. Some authors favour this usage, others ignore it and use the term antigen in an expanded sense synonymous with this definition of immunogen. In this text the problem does not arise because we are only interested in antibody production and we could therefore use either antigen or immunogen as equivalent terms. We will in fact use the former. Unfortunately, the term antigen has also been used to mean any substance capable of specific binding to an antibody, a sense with which immunogen is not synonymous. In fact such a definition includes materials called haptens which while able to bind to antibodies, cannot alone, induce an immune response. We will avoid using the term antigen in this way in this text.

In this Section we are going to describe antigens from two points of view, firstly the range of substances which can act as antigens and secondly their significances as analytes (since in most analyses involving antigen-antibody interactions it is the antigen which is the substance to be measured, and the antibody which is the reagent). Finally some problems regarding the raising of antibodies of particular types will be discussed.

### 3.2.1. Factors contributing to antigenicity

Of major significance is a consideration of the kinds of molecules which can induce antibody formation since the more diverse the range of molecular structures that can act as antigens the wider will be the scope, versatility and range of applications of the techniques.

The most significant factors seem to be:

- the substance being foreign to that body,
- molecular size,
- the monomeric composition of the polymer,
- the rigidity of molecular structure,
- the degradability of the substance.

Naturally a normal human or animal body does not produce antibodies against its own body constituents no matter what their composition, and hence antigenic materials should inevitably be foreign (ie 'non-self') to that body.

Apart from this, perhaps the most important point is that only relatively large molecules are able to induce antibody synthesis. Antigens have a molecular mass exceeding about 1000 and are therefore generally polymers. Thereafter, other factors being equal, the larger the molecule the more effective it is as an antigen.

The other factors listed are also important. Relatively varied, inflexible structures which can be broken down in the body will tend to have greater antigenicity than ones that lack these features, even when they are of relatively large size. For example:

- many polysaccharides have only one type of monomer in the molecule, and are poor antigens, although some with a more varied structure and of similar size are reasonably antigenic. On the other hand, natural proteins always contain a variety of monomeric units in each molecule (often 20 or more) and since most have a $M_r > 1000$ they are generally excellent antigens;

- flexible molecules such as nucleic acids are normally weak antigens despite being huge in many cases;

- inert plastics, some of which are medically useful because their inertness lends them to being placed permanently into the body, are, as a consequence of this inertness non-antigenic.

In practice therefore most antigens of significance are either proteins or heteropolysaccharides, or their conjugates (glycoproteins, lipopolysaccharides etc). Now because protein biochemistry is so important in this field, it would be a good idea to go through the relevant section of one of the recommended texts in the reading list if the field is unfamiliar to you.

**SAQ 3.2a**

From the following list select those materials that you would imagine to be good antigens.

(*i*) the enzyme RNAase with 124 amino acids of 17 different types,

(*ii*) starch – a large polymer of glucose with an $M_r$ value of $10^4$ to $10^8$,

(*iii*) glucose,

(*iv*) polypropylene,

(*v*) the large structural protein collagen, (consisting predominantly of the amino acids, proline, hydroxyproline and glycine),

(*vi*) the peptidoglycan (peptide and carbohydrate) lattice of intact bacterial cell walls.

**SAQ 3.2b**

Which of the following treatments should produce a good immunological response:

(*i*) amylase from a rabbit isolated, purified and re-injected into that rabbit,

(*ii*) amylase from a rabbit isolated, purified and injected into another rabbit,

(*iii*) amylase from a rabbit injected into a horse?

### 3.2.2. The importance of proteins as analytes

The quantitative estimation of various proteins is of great importance in biology and medicine. While some proteins (eg collagen) act as structural materials and are relatively inert, most proteins interact with other molecules and contribute to many vital biological processes, for example:

- enzymes: catalysts of the chemical reactions of the body,

- hormones: usually peptides or polypeptides which act as regulatory materials and circulate in the blood,

- transport agents: both in the blood and in cell membranes,

- defence proteins: especially antibodies.

At the molecular level these substances directly initiate and control many processes and are normal constituents of the healthy body, but their nature, and more often their concentration, may change in disease. In general, with the exception of specific antibodies, identification is not usually sufficient; a quantitative estimation of mass or concentration is often required. Hence the emphasis of this Unit is more on quantitative methods than on qualitative ones, reflecting the fact that most methods used in clinical biochemistry laboratories need to be quantitative.

Raising antibodies to normal body constituents presents problems which we will come back to in Part 4, but nonetheless can be done and has led to the widespread use of these techniques in the quantitation of body proteins. There are, in addition, clinically important antigens which are not normal constituents of the body, and in their case, demonstration of their presence or simple identification may be the only requirement although quantitation may still be useful. The analysis of these antigens is however, carried out principally in microbiology, forensic and blood serology laboratories and therefore receives less attention in this Unit. Among them are microbial antigens – which can be analysed to demonstrate the presence of an infection, or for the identification of microorganisms.

A particularly important example of a normal body constituent for which qualitative identification may be sufficient are the blood group and histocompatibility antigens. These are antigens which are present on cell surfaces and which function to indicate that those cells are from that body (ie they are self-antigens). They are highly variable between individuals; their identification is vital in assessing compatibility in blood transfusion and tissue transplantation and is part of the work of serology laboratories.

A third example is the tumor specific antigens, produced by certain cancer cells. In some cases these compounds are normally present in the embryo but absent in the healthy adult; they reappear when the cells regress to a primitive state on becoming cancerous.

### 3.2.3. The nature and composition of antibodies

(*a*) Types of antibodies

Antibodies circulate in the blood and belong to a group of blood proteins (globulins) which are distinguishable from many other blood and cell proteins by some unusual solubility characteristics. Their membership of this group has meant that antibodies are sometimes called immunoglobulins.

It is possible to separate, eg by electrophoresis, several fractions of immunoglobulins. The molecules in the different fractions have different physiological properties and different roles in the defence of the body. With Ig standing for immunoglobulin these fractions are referred to as IgA, IgD, IgE, IgG, and IgM. These categories are actually groups of molecules, collected together by virtue of similarities in their biological functions, rather than by their binding specificities.

### (*b*) The antigenic determinant

Antigen molecules bind to receptors on the surface of antibody-producing cells (lymphocytes) by small parts of the molecule and not over the entire or even a large part of the surface of the molecule. The binding sites on an antigen are called the antigenic determinants or epitopes and since the lymphocytes will be stimulated only by these, the antibodies produced will only respond to them. The production of distinguishable antibodies therefore requires differences in epitopes.

However even when a single pure antigen is injected, the antibody which is formed, though specific to the antigen, is not a single molecular species and as this is technically a very important point the reasons for it and methods for minimising its consequences will be discussed.

### (*c*) The structure of antibodies

Both quantitatively in the blood and as analytical reagents, the most important antibody fraction is IgG and we will restrict our discussion to this. A simplified diagram of the structure of the IgG molecule is shown in Fig. 3.2a. The two longer ('heavy') polypeptide chains are identical to each other within any one molecule as are the two shorter ('light') chains. The two antigen binding sites are identical and in fact the molecule is symmetrical about the axis (→ ←) in Fig. 3.2a. There is some overall similarity in the structure of all antibody molecules, irrespective of fraction and binding specificity and most of the basic features of IgG shown in Fig. 3.2a are found in the molecules of the other fractions.

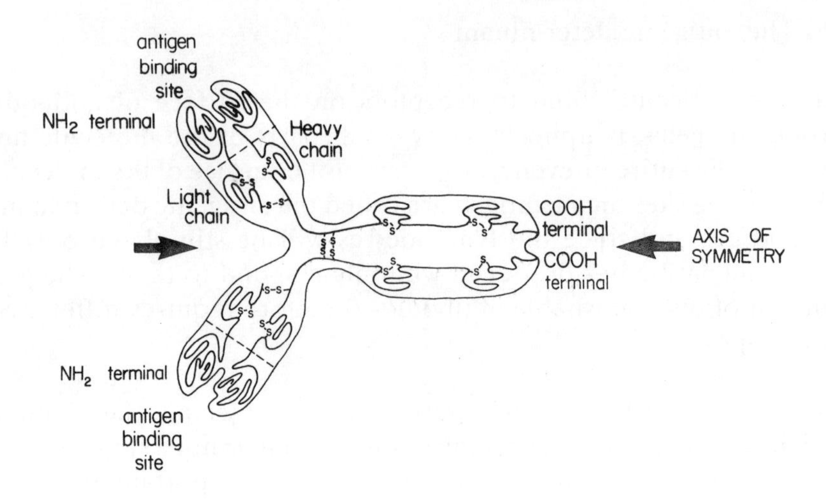


**Fig. 3.2a.** *The structure of an IgG antibody molecule*

Each polypeptide chain is divisible into two sections;

- the variable region at the N-terminal end in which the amino acid sequence is highly variable between individual molecules, and

- the constant region which occupies the rest of the chain, in which the sequence is relatively constant.

The binding sites are formed entirely by the variable regions of the chains. Though identical on any one type of antibody molecule, the antigen binding sites differ between antibody molecules with different binding specificities. It is the amino acid sequences in the contributing variable regions which dictate the binding specificity of the antibody molecule. The so-called constant regions only vary significantly between antibodies of different classes (Fig. 3.2b).

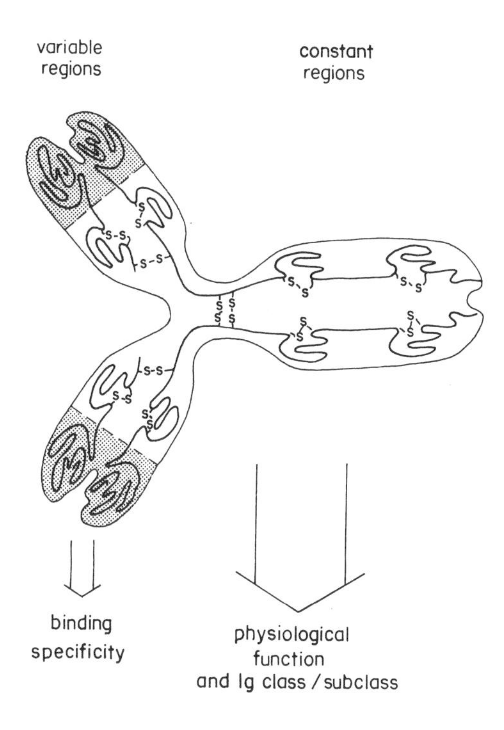


**Fig. 3.2b.** *The main regions of an IgG antibody molecule and their functions*

(*d*) The number and variety of determinants

Problems arise because most antigenic molecules have several binding sites, epitopes, varying from one to about forty, dependent on the size of the molecule (Fig. 3.2c). Basically the larger the antigenic molecule the greater the number of these sites and such antigens are said to be 'multivalent'. In immunology, the valency of an antigen or antibody refers to the number of binding sites per molecule and this is not precisely the same as the conventional use of the term in chemistry. In an antigen-lymphocyte or antigen-antibody reaction there is no requirement that all valencies (ie binding sites) are satisfied; in a particular system or on a particular occasion some may be unoccupied.

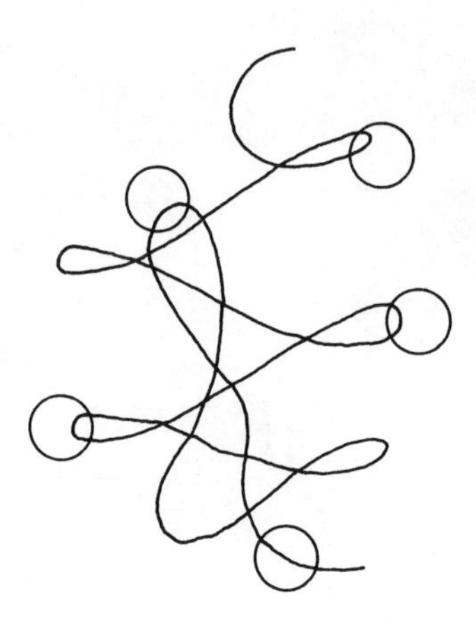

**Fig. 3.2c.** *Schematic diagram showing the presence of antigenic sites (determinants), represented by circles on the surface of a protein molecule*

Π Do you think that all of the binding sites on an antigen will be similar to each other or different?

In fact it will depend on the nature of the antigen. In the case of most polysaçcharides, the polymer is usually made up of only a few (perhaps just one) types of monomeric unit. The sequence is often a repeating one, so that even if there are two types of monomer, they frequently are found to alternate in the polymer chain. Consequently, although there may be many binding sites on each molecule they may in fact be identical or of few different types.

A different situation usually obtains with proteins, since most of them are polymers of 20 or so different amino acid monomer units (insulin has 16) arranged in an essentially non-repeating sequence. As an example the structure of the insulin molecule is given (Fig. 3.2d), and you should bear in mind that this is only a polypeptide ($M_r$ of 5700 for each individual molecule of the physiologically active complex), and not a full sized protein.

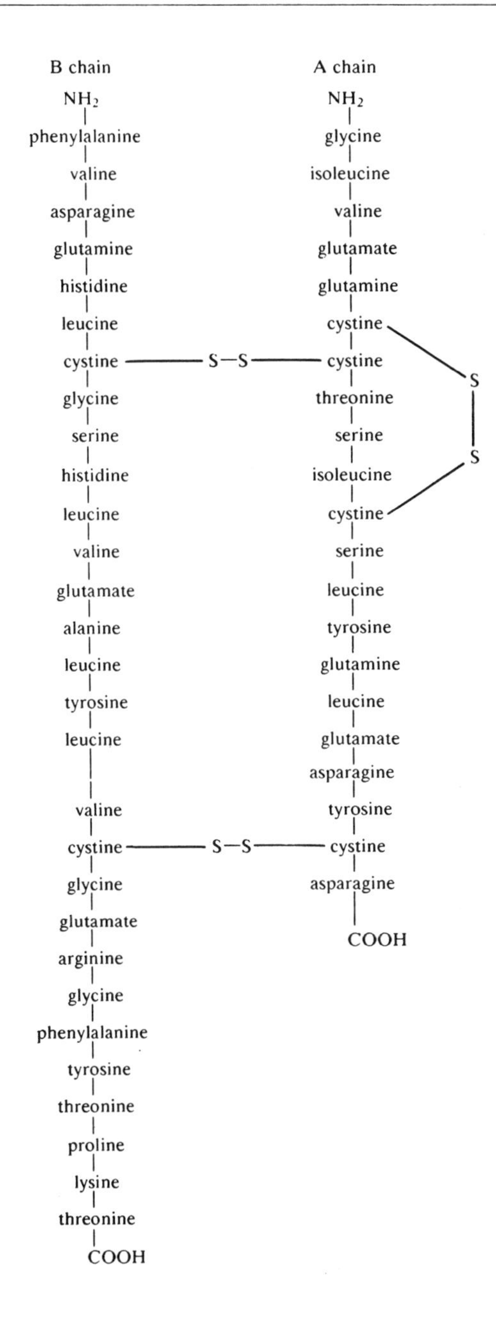


**Fig. 3.2d.** *A representation of the amino acid composition of the human insulin molecule to illustrate the complexity of protein structure*

The sites on a protein antigen are therefore mostly all different and each will be capable of interacting with a lymphocyte to produce antibodies specific for that site. Thus even when a pure protein antigen is injected into an animal, the specific antibody produced is not a single molecular species but a mixed population of molecules all capable of binding the same antigen but at different sites or groups of sites. A further complexity is that in many cases the antigen used will not be a single molecular species in any case, since apart from genuine impurities and contaminants, many biological compounds have a range of related molecular forms which are similar enough to be difficult, time-consuming or expensive to separate (eg the isoenzymes of some enzymes). In such cases an even wider range of antibodies will be produced. Mixtures of antibodies produced as a consequence of these effects are properly called polyclonal antibodies, but are nonetheless frequently referred to in the singular as antibody. The fact that a polyclonal antibody reagent is a complex mixture of reagent molecules of undefined composition which will almost certainly vary from batch to batch, and thus give complex, variable, and ill-defined antigen-antibody reactions, can present problems in analyses with techniques using antibodies.

(*e*) Cross-reactions

The achievement of a good fit in three dimensions accounts for the high degree of specificity of antigen-antibody binding; that specificity is not however absolute.

Cross-reactions can occur, ie a single antibody molecule raised against a pure antigen may also be capable of binding (cross-reacting) with other antigenic substances. Usually these molecules are closely related to the original antigen, and their binding to the antibody is weaker. As we have seen antigens have several binding sites, so cross reactions with the same antibody may be explained, at least partly by supposing that they have some sites in common. In Fig. 3.2e substance B may cross-react with antibody raised against A because it has some binding sites (arrows) in common with A.

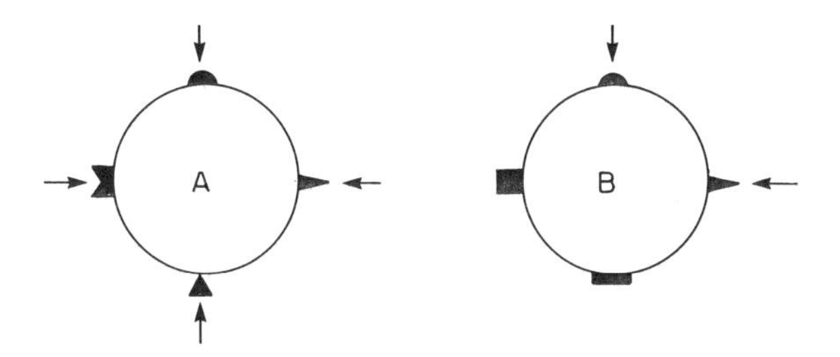


**Fig. 3.2e.** *The development of cross-reactivity due to similarities in antibody binding sites*

This is true so far as it goes, but cross-reactions can also be demonstrated in reactions between antibodies and materials called haptens. These will be discussed more fully later but for the moment regard them as the equivalent of the single site detached from a larger molecule. They are therefore univalent entities and the possibility of cross-reactions between antibodies and different haptens means that specificity is not absolute for a single site either, and two related but slightly different antigenic determinants may be capable of generating and thus binding the same antibody molecule (Fig. 3.2f).

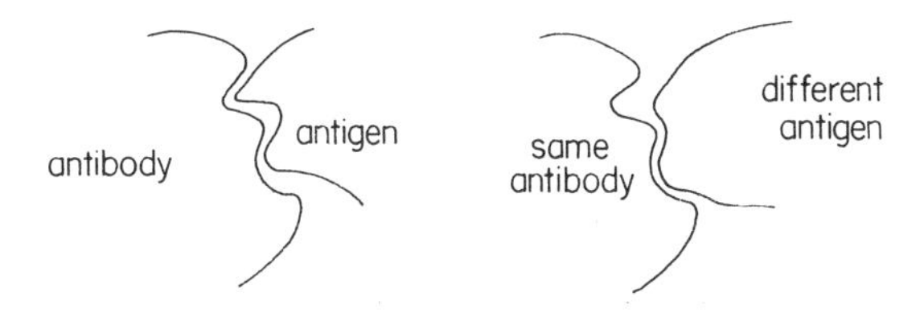


**Fig. 3.2f.** *The development of cross-reactivity due to the fit of different antigens to the antibody binding site resulting in binding to the site*

Conversely slightly different antibody molecules may be capable of binding the same site on an antigen, or binding to the same hapten (Fig. 3.2g). This lack of absolute specificity is due to the ability of:

- molecules other than the original antigen to fit the antibody binding site and bind to it or

- different antibody molecules to bind to the same antigen.

The former may also apply to the lymphocyte-antigen interaction. Different lymphocyte types (clones) may be able to bind and be stimulated by the same site on an antigen, the differences in their metabolism meaning that each clone will then produce slightly different antibody molecules. Ultimately a range of antibodies will result which bind the same antigen site but with differing strengths (affinities).

**Fig. 3.2g.** *The development of cross-reactivity due to the fit of different antibody to the antigen resulting in binding*

(*f*) The antigenicity of antibodies.

Π Since antibodies are large proteins do you think they could themselves be antigenic?

Well in fact they can act as very good antigens and it is important that we consider this further for a while.

There are of course distinct structural differences between the antibodies of different species, so the lymphocytes of a species such as a goat can recognise a guinea-pig immunoglobulin (antibody) as foreign and produce their own antibody against it. Such an anti-immunoglobulin antibody is called a secondary or second antibody.

This goat second antibody will contain molecules directed against species-characteristic determinants on the constant regions of the guinea-pig immunoglobulins to which they will bind irrespective of the binding specificity of the guinea-pig antibody itself.

This facility means that most of the analytical methods using antibody as reagent (Parts 5 and 6) can in principle be applied to the analysis of antibodies themselves. For the purposes of the analysis, the antibody which is being measured as the analyte acts as antigen, and a second antibody raised against it acts as reagent antibody or binding agent.

Second antibody raised against a reagent antibody can also be used as a means of labelling the primary reagent in an analytical procedure (Fig. 3.2h *i*), or in separations, by virtue of the development of sufficient cross-links to produce large aggregates which will precipitate out of solution (Fig. 3.2h *ii*).

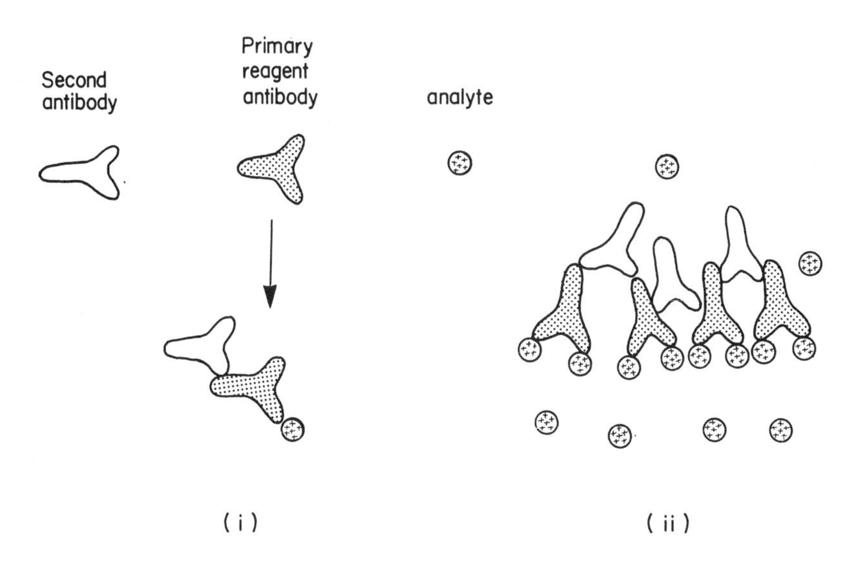


**Fig. 3.2h.** *The use of second antibodies to (i) label a primary reagent antibody and (ii) induce the formation of precipitates by cross-linking*

(*g*) Monoclonal and polyclonal antibodies

Since an individual lymphocyte can produce antibody which is specific to a single type of site on an antigen and all cells produced from it by division (a clone) are the same; it follows that the clone will produce antibody molecules of a single specificity. Such antibodies are called monoclonal antibodies and can, for most purposes, be regarded as consisting of antibody molecules which are absolutely identical to each other. That is they are of a single molecular species, and as exactly equivalent to one of the numerous constituent monoclonal antibodies which go to make up a polyclonal antibody in the blood system. The composition will therefore be defined and unvarying.

Pure monoclonal antibody cannot be obtained simply by immunising an animal for reasons described earlier, but it can now be produced by a laboratory method in which an individual lymphocyte is selected and a clone produced from it by cell culturing techniques. Monoclonal antibodies have considerable potential as analytical reagents despite their cost because of their 'purity' and the possibility of producing them from the cultured clone in considerable quantities.

**SAQ 3.2c**

(*i*) Which of the following can by themselves induce antibody formation:

antigens,
haptens,
immunogens?

(*ii*) What do the following terms mean:

epitope,
antigenic determinant,
valency? ⟶

**SAQ 3.2c (cont.)**

(*iii*) How many binding sites would you expect to find on:

antigens,
haptens?

(*iv*) Which of the following would have a greater diversity of binding sites:

carbohydrates
proteins?

(*v*) Give an alternative name for antibodies.

(*vi*) Distinguish between monoclonal and polyclonal antibodies.

(*vii*) Is it true that the amino acid sequence in the variable region of the antibody polypeptide chain is responsible for its antigenic binding specificity?

(*viii*) Are the antibody binding sites on a single protein antigenic molecule identical?

(*ix*) When an IgG molecule is itself acting as an antigen, would all the determinants on its surface be different?

(*x*) In very simple terms how could antibody capable of binding say, anti-human insulin antibody which has been raised in guinea-pig, be prepared?

(*xi*) Suggest possible bases for the fact that antiserum raised against unpurified chicken ovalbumin (egg white protein) binds both duck ovalbumin (the corresponding protein in a different species), and chicken serum albumin (a different protein from the same species).

**SAQ 3.2c**

### 3.2.4. Forces binding antibody and antigen

An understanding of the forces holding antigen and antibody molecules together is of some importance in these techniques and we will briefly discuss them in this Section.

Covalent bonds are not involved as they are relatively strong with a high energy of activation of formation, and under the mild conditions obtaining in biological systems, enzymes and a source of energy for the new chemical bond would be required. In practice antigen-antibody reactions occur spontaneously without the aid of enzymes and only weaker forces eg, hydrogen bonds, hydrophobic and van der Waals' forces are involved. There are undoubtedly a number of such interactions between a particular determinant on an antigen molecule and the complementary binding site on an antibody molecule as indicated by the arrows in Fig. 3.2i. Though individually weak, these non-covalent interactions can provide collectively a fairly strong bond between the two sites. The strength of this binding as well as its specificity are two of the major merits of antibodies as reagents (see 3.2.5).

These weak non-covalent forces act over short distances only, for a highly specific interaction the two combining sites must complement each other and match very closely (Fig. 3.2i) in order to bring appropriate interacting groups into close proximity. The simplest way to envisage the specificity of antigen-antibody binding is to picture it as a lock and key situation. It is of course the necessity for this close approximation that can lead to the cross-reactivity described in 3.2.3.

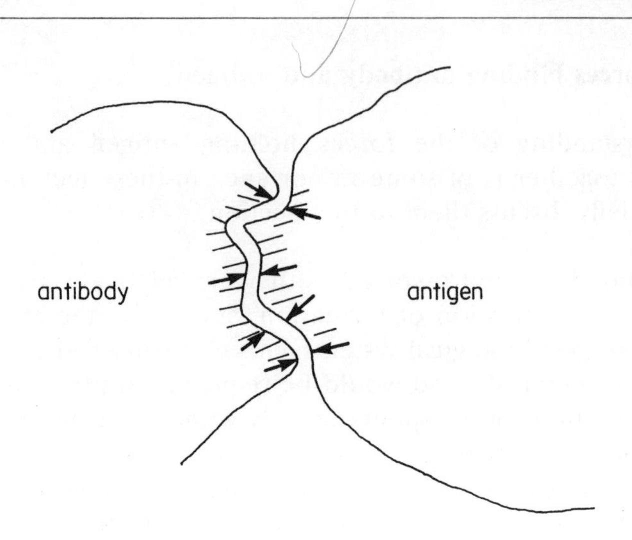


**Fig. 3.2i.** *Diagrammatic representation of the complementary, close fit required of the antigen/antibody interaction in order to generate specificity*

All but the strongest antigen-antibody bindings are fairly easily reversed, eg by adding excess of either reactant or by a change of pH. The influence and control of environmental factors such as pH, and of the nature and concentration of other solutes etc have important implications in the design and execution of the experimental procedures actually employed in these techniques.

The strength of the binding (affinity) is an important factor in determining the sensitivity, reproducibility and specificity of labelled binding methods and therefore a brief explanation of the concept of affinity is given in Section 3.2.5. Though not part of the analysis procedure itself, a determination of the affinities of different batches of binding reagent is a useful preliminary in checking their suitability.

Finally it is perhaps useful to note at this point that one sometimes sees the term 'ligand' as a general name for all materials capable of binding in these reactions, a term which would include antigens, haptens and even antibodies themselves in some situations.

### 3.2.5. Affinity

The strength of the binding reaction is measured by determining the affinity, $K_a$, which is in effect the equilibrium constant for the reversible reaction between antibody ($B$) and ligand (L),

$$B + L \rightleftarrows BL$$

Affinity is given by the equation

$$K_a = \frac{[BL]_e}{[B]_e[L]_e}$$

where $[BL]_e$, $[B]_e$ and $[L]_e$ are the equilibrium concentrations of bound and free antibody molecules and of free ligand respectively. $K_a$ is usually expressed in terms of $dm^3\ mol^{-1}$ (or litre $mol^{-1}$), and typical values for reactions between antibodies and their ligands fall in the range $10^5$ to $10^{12}$ $dm^3\ mol^{-1}$. High affinity antibody of at least $10^8$ to $10^{10}$ $dm^3\ mol^{-1}$ is required for the more sensitive methods. What exactly the experimentally measured value of $K_a$ represents can be problematic, especially for reactions between polyclonal antibody and multivalent antigen, since it reflects a composite value for the individual $K_a$ values of the different molecular constituents of the polyclonal antibody and their different determinants. For this reason, the vaguer term avidity is sometimes employed to describe the strength of binding by a polyclonal antibody.

### Summary

Antibodies are proteins produced by lymphocyte cells in an animal body in response to the presence of antigenic materials of natural origin eg, microorganism cells or their products, or artificially supplied by injection. Antigens will combine with their specific antibodies to form immune complexes and the antibodies can therefore be used as highly specific binding agents in various assay systems.

**Objectives**

You should now be able to:

- define the terms antigen and antibody and briefly describe the result of mixing the two types of material under appropriate conditions;

- recognise that antibodies play an important part in the defence of the body against microbial infections and that they are themselves analytes of medical and general biological interest;

- explain that an antibody appears in the blood only in response to exposure to an antigen (not necessarily of microbial origin) and that it will bind that antigen with a high degree of specificity;

- discuss briefly the structure of antibody molecules and the importance of variations in their structure;

- distinguish between polyclonal and monoclonal antibodies;

- discuss the range of compounds able to act as antigens and the concept of antigenic determinants;

- discuss with examples the clinical importance of being able to measure proteinaceous antigens in low concentration and with high specificity;

- explain the need for and problems of specificity and cross-reactivity;

- describe some features of the antigen/antibody binding reaction.

# 4. The Nature and Production of Binding Agents of Biological Origin

## Overview

In this Part of the Unit we will consider further the characteristics of antibodies and will introduce other important binding agents. The nature of the materials inducing the synthesis or able to bind to these agents and the procedures and related problems for their production are discussed.

Antibodies are the most useful of the binding agents of biological origin and a description of the preparation of polyclonal antibodies and antisera containing them is included in Section 4.3. Monoclonal antibodies are of increasing importance and are covered in Section 4.4 and this part of the Unit then concludes with a brief discussion of the other binding agents of biological origin that have found application in immunoassay techniques.

## 4.1 HAPTENS

When a free antibody molecule binds to an antigen, it binds a small site only, and if it were possible to remove that site from the antigen without the configuration of the site changing, the antibody molecule would bind to the site alone. In principle then antibody molecules should be able to bind to molecules much smaller in size than antigens and can in practice do so. This fact should make antibody-based binding methods of even greater value in analyses, bearing in mind the enormous number of important molecules of small size that different scientists wish to identify or measure.

While the antibody synthesising cells (B lymphocytes) only recognise and bind to a small site on the antigen, the actual induction of antibody synthesis requires stimulation by a relatively large molecule for reasons that we need not consider. Since the production of antibody capable of binding molecules which are too small to be antigens would very considerably extend the usefulness of antibodies as analytical reagents, some aspects of it are described in this Section. The first problem is to produce an antibody which will bind a given analyte of low molecular mass (a hapten), when the analyte itself cannot induce antibody formation.

### (*a*) Formation of antibody against a hapten

While small molecules do not cause an immune response, it is possible to produce antibodies which will bind them by covalently attaching the small molecule to a large one (termed a carrier) which will induce a response. When injected this conjugate will raise antibody molecules, of a polyclonal type, against the complex. The relative molecular mass of haptens which have been used in this way, lies in a range of 150 to 1000 and this includes many compounds of medical and biological importance, such as vitamins, steroids and peptide hormones. The antigenic carrier is a pure protein, foreign to the animal injected, and is frequently albumin.

(*b*) The hapten-carrier conjugate

Covalent linking of the hapten to the carrier introduces a new binding/recognition site on the carrier surface (or modifies an existing one). In the carrier-hapten complex the hapten is therefore attached to the polymeric backbone of the carrier, and serves as a potential binding site for specific B-cell receptors. Its structure is independent of the carrier structure elsewhere. Those antibody molecules specific to the hapten site (Ab7 and Ab8 in Fig. 4.1a) will normally also bind the free hapten, since the bound hapten molecule should have exactly the same structure as the free molecule, except for the very small region by which the bound hapten links to the carrier.

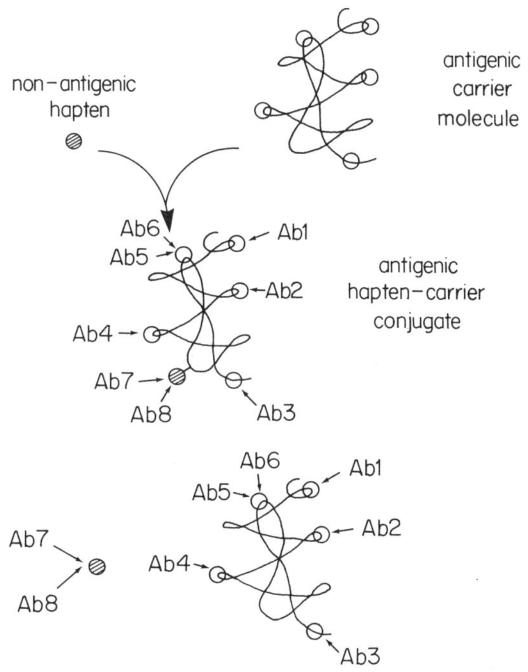

**Fig. 4.1a.** *Production of antibodies to antigenic determinants (circled) on a hapten/carrier conjugate*

Each hapten molecule, either free in solution or attached to a carrier, usually forms just a single binding site (univalent), and in fact the hapten molecule is often so small that it is more or less engulfed by the antibody binding site. In Fig. 4.1a binding sites are circled. The other antibodies (Ab1 to Ab6) bind other sites on the carrier, whether the carrier forms part of the conjugate or not.

Polyclonal antibody raised by immunising an animal with the conjugate will contain antibody molecules capable of binding to all the sites on the complex including that occupied by the hapten. The fact that the 'conjugate solution' used is likely to be a mixture of conjugate, carrier and free hapten molecules should not complicate matters unduly. Similar problems regarding specificity, discussed previously for antigens, apply to haptens also.

Though one hapten molecule is shown linked to a carrier molecule in Fig. 4.1a, in practice when hapten and carrier are reacted, more than one hapten molecule may be inserted on the same carrier molecule (providing several identical sites) and since this will increase antibody yield, techniques are designed to do just this.

(*c*) Linking carrier and hapten

Hapten and carrier can be linked covalently by activation of a group on the hapten so that it will react with one or more of the amino acid residues in the protein carrier. For example, a hapten containing primary amine groups (eg as part of sulphanilic acid), can be activated by diazotisation allowing covalent binding to the phenolic side chain of the amino acid tyrosine (Fig. 4.1b). If reactive groups capable of forming covalent bonds with amino acid residues on the carrier are not present within the hapten, it will be necessary to synthesise an expanded hapten molecule which possesses such reactive groups.

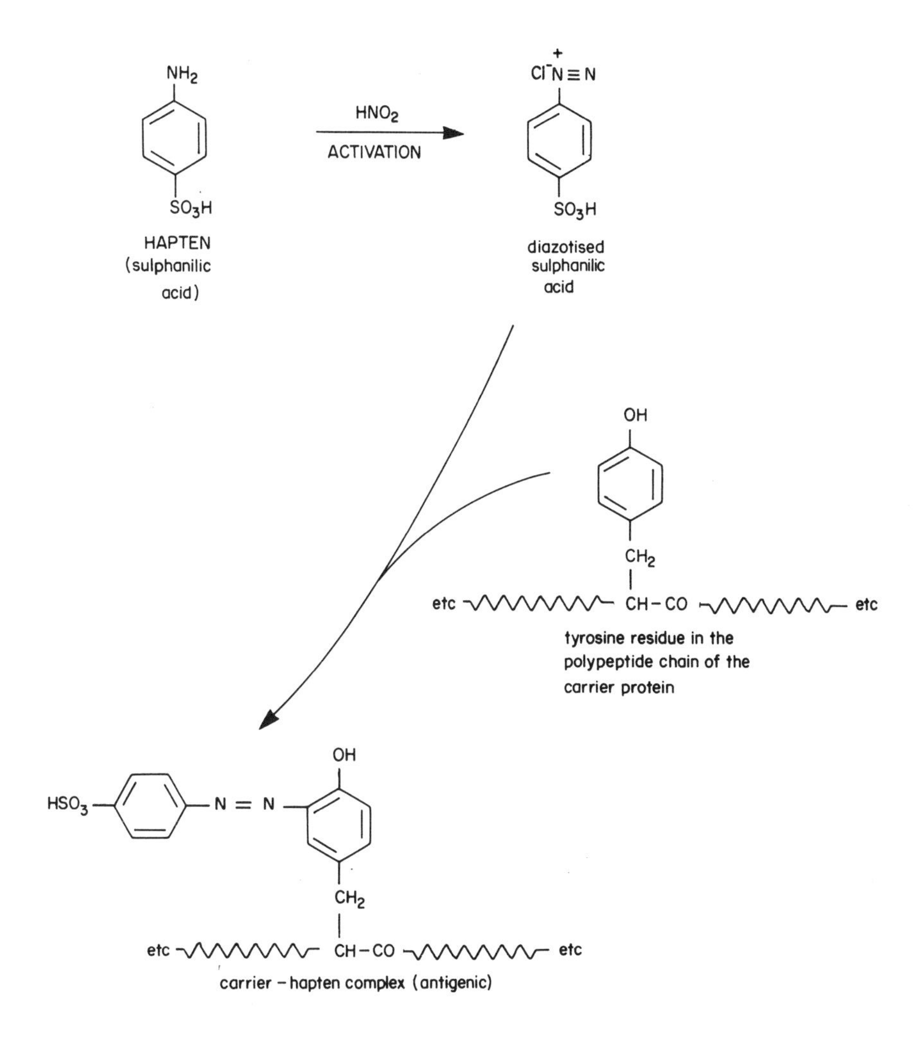


**Fig. 4.1b.** *A typical chemical reaction involved in the formation of a hapten/carrier conjugate*

(*d*) The application of immunological methods to the analysis of haptens.

Most of those immunological methods which monitor the binding reaction itself and which are described in Parts 5 and 6 can be applied directly to haptens just as they are to antigens.

However, haptens are nearly all univalent and immunological methods which rely on the observation of phenomena secondary to the binding reaction and which require some degree of cross-linking, (eg precipitation, light scattering and complement fixation methods, briefly described in 5.1), cannot usually be applied directly to haptens because they require a multivalent ligand. Nevertheless it is possible to adapt even some of these methods to the analysis of haptens by the use of inhibition procedures. For example one in which the hapten competes with a fixed amount of polyvalent antigen for the antibody and so reduces the reaction. The extent of reduction will be related to the amount of hapten present.

(*e*) The importance of haptens as analytes

What types of substance with a molecular mass in the range about 150 to 1000, and which are not therefore directly antigenic, can act as haptens? There is in fact an enormous number of materials which have been used as haptens in this way and their diversity has led to these techniques finding many important applications in fields as disparate as forensic science, clinical diagnostic science and environmental science. Among these compounds are steroids (including the steroid hormones), vitamins, drugs, pesticides and other xenobiotics. Some are difficult to analyse by conventional physicochemical methods because of the structural similarities between related compounds, eg between a drug and its metabolites or precursors, or because of the low concentrations in which they occur, eg vitamins and coenzymes.

The steroid hormones are a good example of the situation in which several important chemically very similar compounds occur in the blood at low concentrations but with the various forms having very different physiological actions. Figure 4.1c gives the structural formulae of three of the oestrogen group of hormones which are of major significance in the development and regulation of female sex organs and characteristics, and in the control of the menstrual cycle. You can readily see their structural similarity and perhaps imagine the difficulty in distinguishing them during their measurement.

Determination of their concentrations in the blood can be used to assess the condition of the glands secreting them, and to monitor the processes they control. While alternative analytical methods were available for some years before immunoassays were developed, in the main they lacked the specificity and low detection limits of the present immunoassay methods. It is not an understatement to say that the development of these highly specific, very sensitive assays for haptens and antigenic hormones has revolutionised some areas of medicine, for example our ability to investigate abnormalities in reproduction. We can identify individual, but structurally related hormones, with relative ease, measure their very low concentrations, and carry out provocative tests to determine whether an organ is responding as it should by secreting the correct hormone at the correct concentration as a result of the stimulus given.

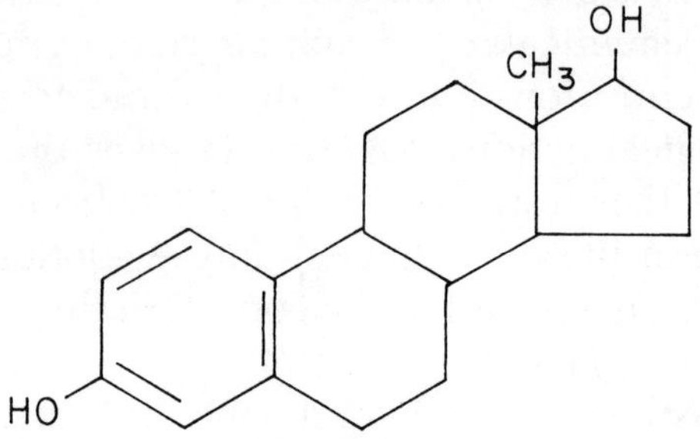

Oestradiol

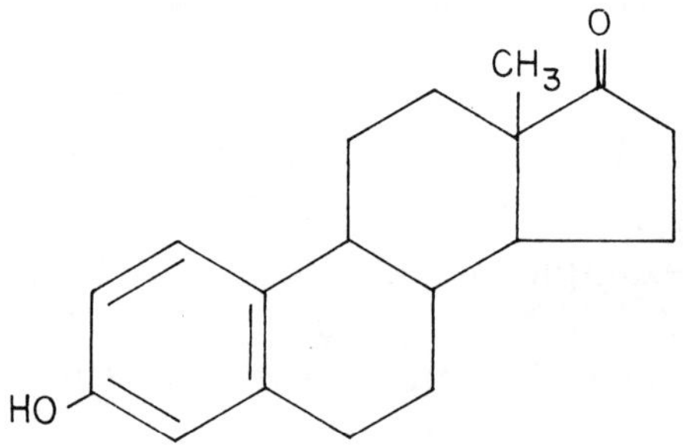

Oestrone

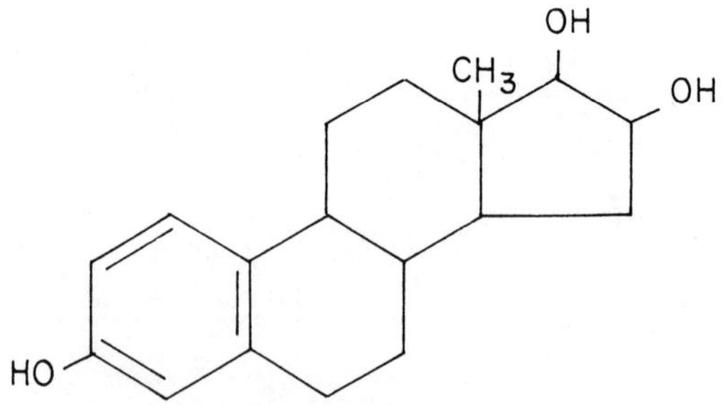

Oestriol

**Fig. 4.1c.** *The structures of three important oestrogen hormones*

## 4.2. LIMITATIONS OF ANTIBODIES AS ANALYTICAL REAGENTS

Antibodies are proteins, and while an indication of their structure has been given earlier (3.2.3c), in reality their structure and chemistry are quite complicated and can be covered only superficially in this Unit. However, some properties are of particular relevance in the present context.

Antibodies, like other proteins, will not withstand heating or extreme pH without loss of their particular three-dimensional structure and their characteristic properties (a process called denaturation). Analytical methods employing antibodies can therefore be carried out only under relatively mild conditions. Similarly the conditions used for any purification or labelling of antibody must be comparatively mild. As proteins, antibodies are potentially food for bacteria and the prevention of bacterial growth in antibody preparations is essential; storage is therefore usually at 4 °C (with the addition of sodium azide as an antimicrobial agent), at −20 °C, or by freeze-drying. These factors have a marked influence on the procedures involved in the production and use of antibodies. Having stated this it is worth noting that compared with many other proteins antibodies are relatively robust.

While accepting that antibodies are extremely useful analytical reagents let us conclude this section by summarising the main problems and limitations in their use.

- Although highly specific, the specificity is not usually absolute and cross-reactions can occur.

- Physiological activity and immunological activity may not match, and this can be important if an investigation of physiological activity is being undertaken.

- Many antibody-antigen reactions are complex and ill-defined, particularly when using polyclonal antibody (3.2.3). This may give rise to interference in the assay (4.3.4).

- Being proteins, antibodies may be denatured by high temperatures, high concentrations of salts and by extremes of pH. They may also be degraded by bacteria and certain enzymes.

- The production of specific antibodies may be protracted and expensive (4.3 and 4.4).

**SAQ 4.2a**

(*i*) Which of the following pairs associate by covalent and which by non-covalent bonds:

antigens and antibodies;
haptens and carrier molecules?

(*ii*) Which of the following should be regarded as a hapten rather than as an antibody:

egg white albumin;
haemoglobin;
thyroxine (a dipeptide hormone)?

(*iii*) About what proportion of the total antibody would tend to be bound, in each case, if a large quantity of hapten was mixed with a small (ie limiting) quantity of the appropriate:

monoclonal antibody;
polyclonal antibody?

(*iv*) Give two of the limitations to the use of antibodies as reagents.

**SAQ 4.2a**

## 4.3. POLYCLONAL ANTIBODY PRODUCTION

When an animal is injected with an antigen, a mixture of antibody molecules is produced. They will all bind the antigen, but do not all bind at the same sites on the antigen or with equal strength (3.2.3). They represent a polyclonal antibody.

### 4.3.1. The use of adjuvant

When producing polyclonal antibody for analytical purposes, injections are usually prepared by mixing the immunising antigen or antigens with an adjuvant, which enhances antibody production in a non-specific manner. The antigen may be in free solution or, if particulate or cellular, in an aqueous suspension. The adjuvant is a mixture of mineral oil and detergent (incomplete adjuvant) or of oil, detergent and a general stimulant of the immune system (complete adjuvant). The mixture of antigen and adjuvant, ready for injection, should form a stable water-in-oil emulsion. This injection procedure with added adjuvant serves both:

- to ensure a slow release of the antigen from the injection site(s), helping to reduce any toxic action the antigen (and adjuvant) may have, and to prolong the exposure of the immune system to the antigen, and

- to stimulate the immune system in a general way so that its response to this or indeed any antigen is improved. In this regard the complete adjuvant is better than the incomplete form.

### 4.3.2. Immunisation programmes

Subcutaneous (under the skin) or intramuscular injection is usually employed since the adjuvant can be poisonous if injected directly into the blood stream (ie intravenously). Each dose is in fact usually split and distributed between several sites on the animal, in the case of rabbits for example between the neck, thighs and sometimes the footpads.

Π Do you remember whether the response to a second dose of the same antigen was unchanged, enhanced or diminished?

Since the antibody response to second, delayed injections is significantly enhanced (3.1.3), the achievement of a high rate of antibody production is promoted by the administration of two carefully spaced sets of injections. Additional ('booster') doses beyond these two do increase the yield but only to a limited and variable extent because there are factors which prevent the lymphocyte clone expanding *ad infinitum*. In practice it is difficult to predict what will happen in an immunisation programme. There may be considerable variation between individual animals even from an inbred strain and to allow for this it is usual to treat a number of animals (say 3 to 6), in the expectation that at least one of them will develop satisfactory antibody production.

The optimum immunisation programme, ie dose size of antigen, frequency and number of injections, time of blood collection etc is difficult to define because of the unpredictable individual variation between animals, and the effect of particular antigens. Immunisation programmes therefore tend to be rather arbitrary and vary somewhat between different laboratories. Some generalisations can be made however:

- antibody production is not related to the mass of the individual dose. Each dose (or total of injections if it is administered at several sites) is usually 10–100 $\mu$g. Doses over 1 mg may actually inhibit antibody formation by immune paralysis;

- in general, the larger the immunogen molecule the higher the yield of antibody. In the case of haptens, a complex set of interrelating factors is important, eg the size and type of the carrier protein and the ratio of hapten to carrier protein molecules.

### 4.3.3. Harvesting antibodies

Antibodies are usually harvested from the blood of the immunised animal and the volume of blood which can safely be taken from an animal obviously depends on its size. Sheep and goats will give much more blood than rabbits and are often used commercially. Nevertheless for many purposes, smaller animals such as rabbits and guinea-pigs will suffice, the level or titre of antibody in the blood often being so high that a single blood sample can provide sufficient reagent for several thousand assays. This is just as well, since very few laboratories have the facilities for keeping larger animals. Choice of animal species is in practice frequently a matter of availability.

The blood taken is allowed to clot and the clear yellow fluid free of red cells (serum) is separated. In this particular case, the serum can be referred to as antiserum, because it contains the required antibodies. In a successful programme the antibodies will contribute about 1% of the total serum protein content. A blood sample will generate about half its volume of serum under normal circumstances, and this is usually sub-divided into aliquots and stored as described earlier (4.2). The antiserum will have to be tested to determine the level of antibody and how strongly and selectively it binds (ie its titre, affinity and specificity). Nowadays screening for blood-borne human diseases (hepatitis, AIDS etc) is also carried out. As with human blood donors animals are unharmed by the blood sampling, and indeed usually by the immunisation processes, so after a suitable interval an appropriate booster injection can be given and further blood collections made from animals identified as giving a good response. It is important in the present climate of public concern to note that animals can be kept and used for these purposes only at premises which are suitably registered, and only by persons holding appropriate Home Office licences.

**SAQ 4.3a**

(*i*) Describe two ways in which complete adjuvant can increase antibody formation.

(*ii*) Name three constituents of a complete adjuvant.

(*iii*) What weight of antigen is usually regarded as maximal for a given series of injections in a small laboratory animal?

(*iv*) What are the usual routes of injection during the production of antibody?

(*v*) Which of the following statements about the quantity of antibody produced in an immunisation programme is/are correct?

- It is directly related to the mass of dose.
- It is influenced by the site of injection.
- It varies with the molecular size of the immunogen.

(*vi*) What is serum, and what does the term antiserum mean?

**SAQ 4.3a**

### 4.3.4. Purity of immunogen and antibody

In some cases, eg labelled ligand assays (5.3), the occurrence of unrelated and unidentified binding reactions, while not desirable, may not affect the reaction of interest provided that is, we are using pure labelled ligand. In theory, assays using labelled ligands, eg radioimmunoassays (5.3 and 6.1), are not affected if other binding reactions occur at the same time, since the one required will be identifiable by its label and only it will be measured. This point will become more apparent when you come to the study of these methods in more detail. What it means in relation to reagent purity is that we do not need a pure antibody preparation as a reagent, and usually antiserum (which will contain low levels of other unidentified polyclonal antibodies) can be used for these methods without any purification at all.

In principle the antigen or hapten (ie prospective analyte) which is used in the preparation of this antiserum does not therefore need to be pure. In practice, however, it can happen that a relatively low level of a contaminant in the immunising injections can produce a disproportionately large amount of its antibody compared to the intended immunogen, because of the complexities and vagaries of the immune response. Then it is possible to end up with quite a high level of specific but unwanted antibody in the preparation which might interfere with even these assays. It is therefore preferable, even for labelled ligand assays, to use as pure an immunogen as possible for the preparation of antiserum.

In other cases, a relatively pure antibody may be highly desirable, necessitating some purification of antibody from antiserum, or the production of specific monoclonal antibody. For example, in labelled antibody assays (5.5), there is an absolute requirement for pure antibody, and both the amount and purity needed are such that polyclonal antibody is a less satisfactory source of reagent than a monoclonal antibody.

Monoclonal antibody is not however suitable for every type of analysis, in particular it does not cross-link well and would not be useful for methods which require extensive cross-linking in order, for instance, to produce precipitates. In techniques such as single radial immunodiffusion it is necessary to measure the distance a sample moves before a visible precipitate develops and in other techniques the immuno-precipitate is quantified by nephelometry. The problem is that few determinant sites are available on monoclonal antibody molecules, and furthermore these may be widely spaced on the molecule making cross-linking difficult.

For the success of methods involving the assessment of the end result by means of light scattering due to particles of precipitated aggregates (5.1), it is important to ensure that no binding reaction occurs other than the one required, but nonetheless cross-linking between various components is necessary. These methods therefore need an immunising material which is as pure as possible, and some purification of the required antibody from the antiserum (often by affinity chromatography with the appropriate ligand linked to the matrix as the stationary phase) is often employed. The high specificity of the affinity chromatography system would remove antibodies of other specificities as they would not be retained on the column. Antibodies purified from antiserum in this sort of way are called monospecific antibodies, and the fact that they will still be a polyclonal mixture will ensure that they are capable of entering into cross-linking reactions.

Another problem with monoclonal antibodies is that they are more likely to undergo fortuitous cross-reaction with unrelated molecules, ie those having similar determinants to the original immunogen but otherwise being dissimilar, due to the reaction involving relatively few binding sites. Finally the binding reactions of different monoclonal antibodies to the same antigen may have very different relationships to the physiological action of the antigen. This will depend on how closely each determinant on the original immunogen is associated with the site of physiological action. For example, a monoclonal antibody binding to a determinant far removed from the active site of an enzyme antigen will not affect the enzyme activity. Conversely, a clone may be selected which produces antibodies which bind to a determinant that is much closer to the enzyme active site and which will therefore have great enzyme blocking activity on binding. In an assay the latter are more likely to give a true measure of the physiologically active (in this case enzymatically active) analyte molecules in the sample. The former may bind equally well to molecules with such structural defects that they have little or no enzyme activity, but this would not be apparent from a study of, or using a method based upon, their immunological response. Such structural changes may not affect observed enzyme activity and immunological responses to the same extent.

## ...LONAL ANTIBODIES

...ntibodies can now be produced in the laboratory in ...nt quantities and while they have some extremely useful properties and characteristics they are unlikely to replace polyclonal antibodies entirely as analytical reagents for financial reasons as well as those described above (4.3.4).

**Monoclonal antibody production**

The antibodies in an antiserum are inevitably polyclonal. As discussed earlier it is not really possible to obtain pure monoclonal antibody simply by immunising an animal and extracting the antiserum no matter how pure the antigen. However methods are available for monoclonal antibody production which involve selection of a suitable clone of antibody producing cells, and growth of this outside the body in a suitable medium (a process of cell or tissue culture). It is important that we discuss the procedure involved briefly.

In some cases of the rare disease, myelomatosis, monoclonal antibody is produced in large quantities without apparent immunological stimulus. What has happened is that a lymphocyte has been transformed into a cancer cell and has then divided repeatedly forming a single clone of cancer cells (a myeloma), which secretes large amounts of a monoclonal antibody, referred to as myeloma protein. These cells differ from normal lymphocytes in that they divide without immunogenic stimulus and are not subject to normal cell regulation processes; they have though, retained their antibody production capacity.

This transformation can also be initiated experimentally outside the body enabling the production of a monoclonal antibody but in such a way that the binding specificity of the antibody can be selected.

Normal lymphocytes, like other normal cells, cannot live for very long outside the body even under the most favourable tissue culture conditions. Cancer cells on the other hand can live and divide indefinitely in tissue culture; they are effectively immortal. So if a normal lymphocyte is fused experimentally with a cancer cell, the resulting hybrid cell retains some of the properties of the cancer cell, in particular the important property of living and dividing indefinitely to produce a clone. The hybrid also has some of the properties of the parent lymphocyte, the most useful of which is its ability to secrete antibody. The combination of these properties means that significant amounts of monoclonal antibody will be produced over a long period of time and can be collected for experimental purposes (Fig. 4.4a).

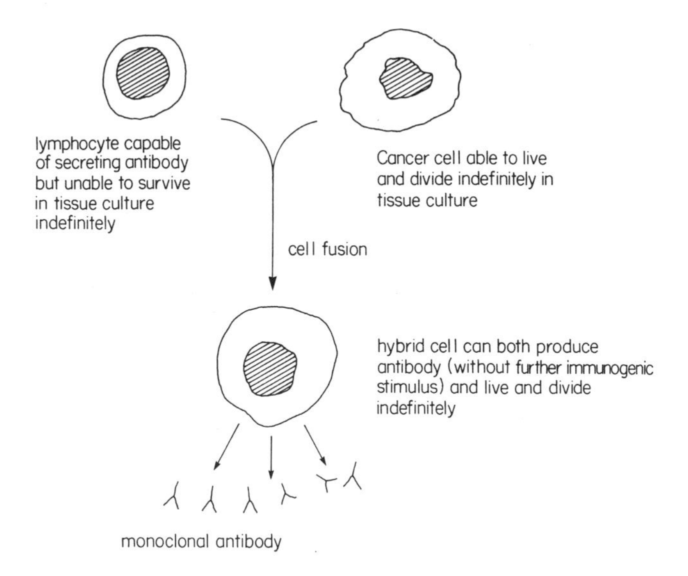


**Fig. 4.4a.** *The production of hybrid cells for the manufacture of monoclonal antibody*

To obtain antibody of the required specificity, lymphocytes are taken without selection, from an animal previously immunised with the required antigen. Even after the immunisation, only a small proportion of these lymphocytes will be able to produce the required antibody, but hybrid cells are formed from all the lymphocytes in a non-selective way. The hybrid cells are isolated by growth in a medium toxic to unfused myeloma cells, the unfused lymphocytes failing to grow in any medium. The surviving hybrids are then tested, and those producing the required antibody identified and grown into separate clones. The clone is grown in tissue culture, allowing the large scale production of monoclonal antibodies (in tens of grams per batch); of course these are free from contamination with other immunoglobulins. An alternative method of large scale production is to introduce the selected clone into the peritoneal cavity of an animal of the same species and strain, and after growth of the cells into an ascites tumour peritoneal fluid can be aspirated as a source of the antibody. Not surprisingly the prevailing attitude to the use of experimental animals in laboratories has made this approach less popular.

**SAQ 4.4a**

If a conjugate of the steroid hormone progesterone and a carrier is used to immunise a mouse and the lymphocytes removed, fused to form hybrids which are grown into clones, will the antibodies produced by the different clones be identical?

**SAQ 4.4b**

Answer each of the following questions with the term 'monoclonal' or 'polyclonal'.

(*i*) Which type of antibody requires the more complex initial preparation and is in general more difficult to produce?

(*ii*) For the production of which antibody is the purity of the initial immunogen of greater importance?

(*iii*) Which antibody has less defined and more variable composition?

(*iv*) Which antibody is not absolutely specific to the immunogen?

(*v*) Sustained large scale production is possible for which antibody?

(*vi*) Which antibody is more suitable for precipitation reactions?

**SAQ 4.4c**

What is unusual about the protein metabolism of cells involved in the disease, myelomatosis, and in what way does the cell division and survival characteristics of these cells differ from those of lymphocytes?

## 4.5. OTHER BINDING AGENTS OF BIOLOGICAL ORIGIN

As analytical reagents antibodies are much the most useful of those produced biologically but they are not the only ones. Other selective biological binding agents include the serum transport proteins, described in 4.5.1, and cell receptors, described in 4.5.2. Both have the merit that they are present in the body all the time and their synthesis does not have to be induced as it does for specific antibodies. However they are available for relatively few ligands and it is not usually possible to produce them in large quantities.

### 4.5.1. Serum transport proteins

Here the binding agent is a protein used by the body to transport a specific ligand. In this case the binding of the ligand to the protein binding site does not result in a physiological response; the body evolved this process to enable transportation in the aqueous blood system of compounds that otherwise are not water-soluble, or are toxic, or are unstable, and as a means of keeping a reserve of physiologically inactive material in an equilibrium with the active free pool. Examples include the binding of fatty acids to serum albumin, steroids to glycoproteins, and the binding of hormones such as thyroxine to specific carriers, in this case thyroxine-binding globulin. In each of these examples the affinity of the protein-ligand interaction is of the same order of magnitude as that observed in antibody-immunogen interactions with values (the meaning of which is explained in 3.2.5) of typically $10^8$ to $10^{10}$ $dm^3$ $mol^{-1}$.

The binding is non-covalent and probably involves similar forces to those in antibody-immunogen interactions, however on the whole, the binding of ligands to their transport proteins in less specific than binding to antibodies.

### 4.5.2. Receptors

Receptors are proteins located in the surface membranes of cells. *In vivo*, the binding of the ligand to its receptor initiates a particular physiological response through a series of biochemical reactions, the details of which need not concern us, but are for example the bases of the action of proteinaceous hormones, such as insulin, in the body. In this case the ligands are hormones (which act as chemical messengers and co-ordinators regulating the activity of selected tissues distant from their origin), and they are carried in the blood so that all cells are exposed to them. However, only some cells respond to a given hormone and these are the ones with suitable receptors for those hormone molecules. Herein lies the basis of the selective action of hormones on different tissues of the body.

We need also to appreciate that medicinal chemists have (sometimes inadvertantly) synthesised molecules which have high affinity for certain receptors and these molecules mimic the naturally occurring ligands in initiating the physiological response. Such molecules are generally called agonists, although if their binding to the receptors fails to induce a response and prevents the naturally occurring ligand from binding and initiating the response, they are called antagonists.

In some cases such compounds have proved to be of considerable medical value, eg in the modification of the activity of the so-called $\beta$-adrenergic receptors found on the surface of many cell types and normally controlled by hormones of the adrenaline group. Detailed studies have shown that an amine-containing side chain is required for binding to the receptor and a catechol-like ring for initiation of metabolic activities. Thus by modification of these structures, selective action can be obtained. Compounds such as practolol act as antagonists but only for the $\beta$-receptors of heart muscle, and provide the so-called beta-blockers used in treatment of heart conditions. On the other hand terbutaline is a selective agonist for bronchial smooth muscle and is used in the treatment of asthma.

We can utilise the binding interactions between the receptors and their ligand to assay the ligand in two ways; by bioassays using intact tissue (see Part 1), or by binding assays using tissue extracts. One theoretical advantage is that the binding or response measured represents the physiological potency of the analyte and the use of receptors in either approach should therefore more truly measure the functional activity of the analyte. The point has been made elsewhere that a potential defect in assays using immunological principles is that defective and possibly physiologically inactive molecules may still give a typical immunological response if the appropriate part(s) of the molecule are intact.

You will have seen in the description of bioassays (Part 1) that we can isolate tissues rich in these receptors and monitor the physiological response resulting from the ligand binding to them. This response is proportional to the binding of the ligand to the receptor, and the exposure of the isolated tissue to solutions of known concentration enables us to construct dose-response curves. From these we can estimate the ligand concentration in solutions of unknown concentration.

The alternative approach, important in the present context, is to isolate the receptor protein from the target tissue and to study directly the ligand binding to these receptors. The theory of these receptor assays is identical to the labelled ligand assays described in Section 5.3 and in Part 6. The use of tissue extracts in this way has, however, the disadvantages they they are not always available ready prepared and preparation may require access to fresh tissues and extracts. They are generally less stable than antibodies and furthermore they may contain receptors of other binding specificities. In this regard they are analogous to antiserum. Techniques using isolated receptors are applicable to a restricted though important range of analytes, particularly steroid and protein hormones, and to some drugs and their active metabolic derivatives.

Π Can you think of another way in which the receptor/ligand interaction can be used in quantitative measurement? One important application is really the reverse of the situation described above.

Well the system can just as easily be used to measure the number or functional activity of receptor sites in a given extract as it can the level of binding ligand, and this is quite important since many diseases result from, or in, receptor defects. For instance although the commonest cause of *diabetes mellitus* is a reduction in insulin output by the synthesising glands of the body, other causes do exist such as the production of defective molecules, the neutralisation of insulin molecules by circulating antibodies and the failure of target cells to respond to insulin because of receptor defects or reduced number.

**SAQ 4.5a**

Which of the following statements are true?

(*i*) Both biological assays and receptor assays utilise binding by cell receptors.

(*ii*) Binding by cell receptors and serum transport proteins is generally about as specific as that by antibodies.

(*iii*) Specific serum transport proteins are available for the analysis of relatively few substances.

(*iv*) The presence of the same transport protein in the blood sample for analysis as is used as the binding reagent can complicate the analysis.

**SAQ 4.5b**

Define the terms:

(*i*) serum transport protein;

(*ii*) receptor;

(*iii*) agonist;

**SAQ 4.5c**

In the binding reaction between each of the following pairs of reactants, indicate which component would conventionally be regarded as the ligand.

(*i*) triiodothyronine/anti-triiodothyronine antibody;

(*ii*) guinea-pig anti-human insulin antibody/ human insulin;

(*iii*) guinea-pig antibody/goat anti-guinea-pig Ig antibody;

(*iv*) thyroxine-binding globulin/thyroxine.

**Summary**

Haptens are low molecular mass materials which are not usually able to induce antibody production although coupling of the hapten to a protein carrier molecule will allow them to induce antibody synthesis. It is important to be able to use antibody-based techniques to analyse haptens because of the vast range of compounds of biological and medical significance which are in the molecular mass range of haptens rather than antigens.

Polyclonal antibodies are poorly defined and variable mixtures of related molecules which for historical and economic reasons are the most commonly used types. Techniques for their production using stimulants (adjuvant) and live animals are therefore important. More recently laboratory based techniques have been developed for the large scale production of pure (monoclonal) antibodies using clones of isolated lymphocytes.

Serum transport proteins and cell membrane receptors have also been used as binding agents to analyse the ligand, or conversely appropriate ligands can be used to measure the transport protein or receptor level.

**Objectives**

You should now be able to:

- define the term hapten and discuss the importance of being able to measure them;
- discuss the necessity for being able to produce a hapten/carrier conjugate and the techniques for so doing;
- understand that since haptens are univalent, some limitations on immunological methods using them exist;
- explain the limitations of antibodies as reagents, including some problems originating from the relative instability of antibody proteins;

- describe immunological procedures for the production of polyclonal antibodies including the use of adjuvants;

- discuss the differences between polyclonal and monoclonal antibodies and the value of the latter in immunoassays;

- outline the techniques for the production of monoclonal antibodies;

- describe the nature of serum transport proteins and cell membrane receptors and the importance of being able to analyse them, or their corresponding ligands;

- recognise that one advantage of the investigation of transport proteins and receptors is that the binding reaction represents the actual physiological function of the analyte in contrast to many antigen/antibody systems;

- realise that techniques using transport proteins and receptors are available for relatively few analytes and that the isolation of the latter may require specialised and difficult techniques.

# 5. Applications of Biological Binding Agents in Quantitative Analysis

## Overview

This Part of the bioassay unit will consider the range of procedures in which biological binding agents have been used. Included are simple discussions of methods involving changes secondary to binding and the general principles of assays using labelled ligands. A brief survey of the main variations in the labelled ligand approach is given and discussed. This Part concludes with a description of the fundamentals of assays using labelled antibodies as reagents.

## Introduction

As we have seen, antibodies are highly specific binding agents and can be produced to bind a very wide range of compounds, whether naturally occurring or not. They and, to a lesser extent, the other biological binding agents described in Section 4.5, can be used to separate, identify and quantitate the substances they bind, ie their ligands. While the majority of methods performed in laboratories are for the estimation of ligand using the binder (antibody/transport protein/receptor) as reagent (Fig. 5.0a (*i*)), procedures have been devised to detect or quantitate the binding agent, or rather its binding capacity. These assays for antibody etc are in practice carried out with the target analyte antibody (binder) now acting as a ligand, ie antigen, to a secondary (reagent) antibody (Fig. 5.0a (*ii*)); the original ligand may or may not be involved in the assay. In this Unit we will only describe and discuss procedures for the analysis of ligands, which we will refer to as analytes.

(*i*) Ligand Assay

Ligand $\xrightarrow[\text{by}]{\text{assayed}}$ Suitable binding agent (reagent)

(*ii*) Binding Agent Assay

Binding agent (acting as antigenic ligand) $\xrightarrow[\text{by}]{\text{assayed}}$ Antibody to binding agent (reagent)

**Fig. 5.0a.** *Systems for the assay of (i) ligands, and (ii) binding agents*

In order to use a binding agent to identify or quantitate its ligand, there must be some means of detecting or measuring the extent of the binding reaction. This can be done in several different ways, varying in sensitivity, versatility, principle involved and in suitability for particular systems, but the majority can be divided into:

- those in which the primary binding reaction is monitored directly using a label, which we will call labelled binding methods, and

- those in which a change is monitored which can be regarded as secondary and a consequence of the primary binding reaction. These methods have not found as wide an application in chemical analyses as alternative methods and so will not be discussed at length. However for the sake of completeness a short summary is given in Section 5.1.

## 5.1. METHODS INVOLVING CHANGES SECONDARY TO BINDING

These methods mainly involve reactions between antigens and antibodies rather than other binding agents. The reactions result in secondary changes which can be observed with the naked eye or by suitable optical instruments. We can usefully divide the methods into those involving measurement of (*a*) aggregates, (*b*) larger agglutinates or (*c*) cell lysis.

(*a*) Direct detection or measurement of large molecular aggregates formed as a result of cross-linking reactions between soluble antibody and soluble antigen is a common approach. The aggregates are measured either:

- while still in suspension by using techniques (turbidimetry or nephelometry) based on light scattering effects, or
- after a visible precipitation.

In recent years immunoprecipitation reactions have assumed a significant role in measurement of the many important human proteins which are present at low concentration in complex mixtures containing many other proteins. The human blood proteins caeruloplasmin (a copper-containing oxidase at $<500$ mg dm$^{-3}$) and $\alpha_1$-orosomucoid (a steroid binding protein at $<1400$ mg dm$^{-3}$) are good examples of proteins where the specificity of the immunological reaction is of great significance in their measurement.

The principle of the assay is straightforward enough, optically clear antiserum containing antibody specific to the target protein is mixed with the sample and the resulting precipitate quantified while still in suspension by a light scattering measurement. The techniques have been automated, and laser light sources and forward light scattering nephelometry mean that low detection limits and a precision of $<8\%$ is commonly achieved. Another useful point is that the automatic centrifugal fast analysers, that many laboratories possess for other purposes, can be used in these assays. The time required for assay is relatively short ($<30$ minutes) and can be made even shorter ($<5$ minutes) if the rate of formation (ie kinetic study) of the precipitate rather than the maximum quantity is determined.

Precipitation forms the basis of many important qualitative tests often involving diffusion in a semi-solid gel. These methods are also available for quantitative analysis either by a direct measurement of the mass of the precipitate or, more usually, by measurement of the position of the precipitate formed in a gel following diffusion or electrophoresis of a reactant. For example, if anti-human albumin is dissolved in warm agar and a thin layer of this allowed to solidify on a glass plate, small holes (or wells) can be cut in the agar and filled with biological samples or standard solutions of albumin. The proteins then diffuse from the wells and the antigen and antibody react to form a visible precipitate in the agar. A useful feature of this precipitation reaction is that it only occurs when a certain ratio of antigen and antibody concentration is reached (the equivalence point).

Π Since the antibody concentration is fixed, what measurable parameter do you think will change as the concentration of albumin in the well increases?

As the albumin diffuses out of, and away from, the well its concentration will fall and the higher the initial concentration the further the sample must diffuse before the equivalence point is reached, thus it is the ring diameter or area that is a reflection of sample antigen concentration. Calibration curves can therefore be produced and the concentration in experimental samples measured by comparison.

While the methods have certainly been used in biochemistry, and in the detection and measurement of soluble antigens and antibodies in microbiological and haematological analyses, they suffer from relatively poor detection limits (10–50 mg $dm^{-3}$) and sensitivity. Accuracy of measurement, the length of time required for precipitation to occur, difficulty in automation and a wide range of potential problems arising from other immunologically active molecules in complex biological specimens are additional concerns.

(*b*) Direct detection of a visible agglutination (clumping) of cells or particles resulting from a cross-linking reaction between soluble antibody and cell or particle bound antigen. It is also possible for the antigen to be in solution and the antibody surface bound.

Perhaps the classic example of the use of this technique is the identification of some blood groups by methods in which, for instance, human red blood cells of type A can be identified because they will be cross-linked to form an agglutinated mass by antibodies of type A. However, the basic method is made more versatile by the fact that many soluble antigens will spontaneously attach, or can be chemically attached, to cells or even to artificial particles such as latex. These all act as carriers for the antigens and will agglutinate when the appropriate antibody is present at a suitable concentration.

A further variant of the technique is agglutination inhibition in which the experimental sample is given a preliminary incubation with the antibody so that binding by the sample antigen reduces the effective antibody concentration. The greater the sample antigen concentration the greater the extent of this reduction and the lower the final agglutination response.

These techniques have the merits of low detection limits ($< 10\ \mu g\ dm^{-3}$) and considerable versatility because of the wide range of substances that can be attached to various particles and cells. Difficulties can however be experienced in determining the degree of response, and complex biological specimens can produce a range of interfering effects.

(*c*) Direct detection of a visible rupture of cells resulting from complement fixation (see Glossary) to a cell-bound antibody/antigen complex.

While the mechanism involved in complement fixation reactions is actually quite complex, in principle what happens is that when antigens and antibodies combine, they can activate some members of a group of proteins called complement. These develop enzymic activity and in turn activate other members of the group giving an amplification cascade, which can result in the end in the rupture of cells to which the complexes attach.

In the methods commonly employed, complement from animal serum is consumed by an *in vitro* reaction between an antibody and the target antigen. When red cells are added subsequently, the extent of lysis is a measure of the amount of complement left.

Π When the concentration of antigen in a sample rises would you expect the extent of red cell lysis to rise or fall?

In fact lysis will fall since the immune complex will have consumed more of the complement so that less is available for inter-action with the cells.

The methods have the advantage of a readily identifiable end-point (red cell lysis) and quite low detection limits ($< 10\ \mu g\ dm^{-3}$) due to the amplification by the complement cascade. Difficulties in standardisation of the procedure (the variability in response of different batches of red cells being a particular problem), often lead to difficulties in quantitation, and the methods are more frequently used for microbial (especially viral) identification than chemical analysis.

**SAQ 5.1a** List four methods by which antigen/antibody reactions can be detected.

## 5.2. PRINCIPLES OF LABELLED BINDING METHODS

In these methods a label is introduced to monitor the primary binding reaction. Either reactant (ligand or binder), can be labelled and it is generally the uptake of label into the ligand/binder complex that is detected or measured. The procedure is in principle quantifiable, ie the extent of the binding reaction can be determined by measurement of the label in a suitable way.

Quantitative labelled binding assays are extremely important in chemical analysis and measurement. They provide not only the specificity characteristic of the binding by antibodies and other biological binding agents, but also extreme sensitivity, since it is the label itself which is measured rather than reactants or products. This gives one the freedom to choose labels with particular merits for a given application, but labels which can be measured at very low levels, eg radioisotopes, are frequently employed. Measurement systems for radioisotopes are such that accurate estimations of analytes at concentrations down to pmol $dm^{-3}$ ($10^{-12}$ mol $dm^{-3}$) or sometimes lower are possible. They have enabled the estimation of a large number of substances of medical, forensic, veterinary, environmental and general biological significance for which previously no satisfactory methods existed. Their introduction has proved a major advance in biochemical analysis and we will discuss them in some detail in Parts 5 and 6 of this Unit. Labelled binding reactions also form the basis of a number of important tests, mainly qualitative or semi-quantitative, used in microbiology, haematology, histology and other laboratories.

In this Part we will discuss the principles of the use of labels in detecting or measuring the primary binding reaction, in order to set the scene for a discussion of the technical details of the assays in Part 6.

## 5.3. LABELLED LIGAND ASSAYS

Assays involving the use of a labelled ligand can be performed in a large number of ways. We will begin by discussing a relatively simple and fairly common basic method.

In a typical labelled ligand assay for the estimation of ligand concentration each tube would contain:

- a sample for analysis or a standard containing a variable amount of unlabelled ligand,

- a fixed amount of labelled ligand, and

- a fixed and limiting amount of binder.

The amount of binder needs to be relatively low so that it is insufficient to bind all the ligand molecules. The binder accepts labelled and unlabelled ligand molecules randomly and indiscriminately, but it cannot bind them all (such a method is said to be a limited reagent method). The labelled and unlabelled ligand molecules are said to compete for the binder in a random process involving equals, so that the binding of one ligand molecule necessarily excludes another.

Fig. 5.3a shows that competition occurs at the binding site on the surface of the protein for ligand or labelled ligand only. Other potential ligands (X and Y) have shapes which are incompatible with the binding site and do not participate in the competition. This is the origin of the great specificity of assays employing specific binding sites.

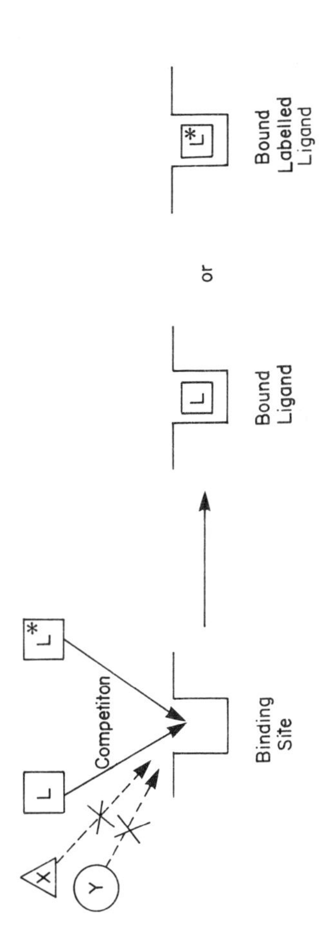


**Fig. 5.3a.** *Competition at the binding site on the surface of the protein for ligand (L) or labelled ligand (L*) only*

In Fig. 5.3b, L is the unlabelled ligand, ie the analyte we are trying to quantify. It competes with a fixed amount of labelled ligand, L*, for a fixed and limiting (low concentration) amount of binder, B. At the end of the reaction, there is a mixture of labelled (BL*) and unlabelled (BL) bound ligand in the complex, and of labelled (L*) and unlabelled (L) free ligand. A very small amount of free binder may also remain (B). If necessary the bound label (ie in practice the total complex, BL + BL*) is then separated from the free label (ie total free ligand, L + L*) and one or the other measured. One potential advantage of the use of a label is that a separation of labelled bound ligand (BL*) from unbound labelled ligand (L*) may not be required if the properties of this label change on binding.

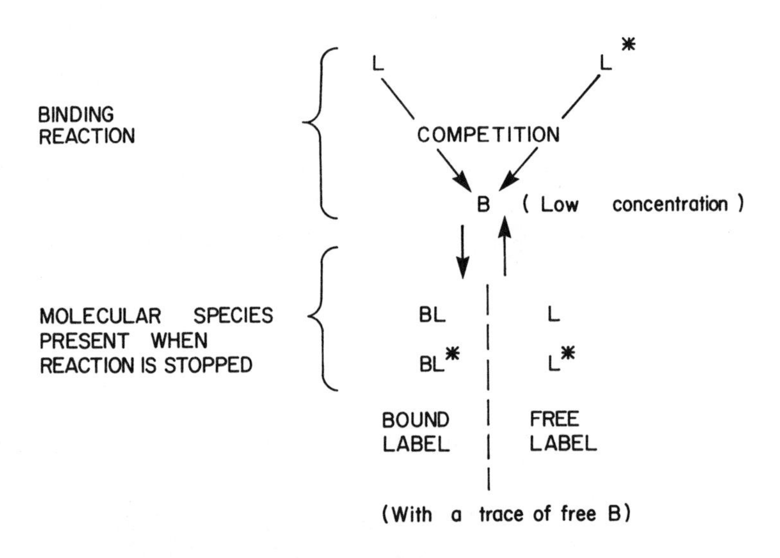


**Fig. 5.3b.** *Schematic diagram of an assay using labelled ligand and limited antibody. L is unlabelled ligand, L* is labelled ligand, B is a binding agent for L and L**

Fig. 5.3c gives a numerical illustration of the principles of the assay using 24 target analyte molecules (open circles), and six labelled molecules (solid circles). The numbers are of course hypothetical and the reactions shown assume that the binder and ligand molecules react on a 1:1 basis (in practice unlikely), and that labelled and unlabelled ligand are bound indiscriminately. The proportion of binder remaining unbound will depend on the affinity and concentrations of reactants but in practice it is going to be minute and could even be ignored in this simple example. The final figures (8:2, and 16:4) show that the ratio of unlabelled to labelled ligand molecules (4:1) is the same in each fraction, ie the labelled ligand has distributed itself between the free and bound state in the same ratio as in the original mixture. It is the case then that the distribution of label is a good reflection of the concentration of unlabelled molecules in the original sample.

Ask yourself what sort of relationship you think exists between the final amount of label in say the bound fraction, and the original concentration of analyte (ie unlabelled ligand) in the sample and what would be the effect of a change in the latter. Well if we increase the amount of unlabelled ligand, it should be obvious from Fig. 5.3c that if we have a fixed amount of labelled ligand, the amount of label finally bound in the complex will decrease (and the final free label increase), because there will be an increased competition for the binder from the greater number of unlabelled ligand molecules (Fig. 5.3d(*i*)). There is therefore an inverse relationship between bound label and original analyte concentration, though not a simple linear one. It is possible to set up a calibration curve using known concentrations of the unlabelled ligand (L) and fixed concentrations of L* and B. As the concentration of L rises the proportion of L* that binds to B will fall, ie the Bound/Free ratio of label will decrease (Fig. 5.3d(*ii*)). Then, when samples which have an unknown concentration of target ligand are incubated with the same concentrations of L* and B, a distribution of label into bound and free fractions will occur. After measurement of this, the concentration of L can be obtained by interpolation from the calibration curve.

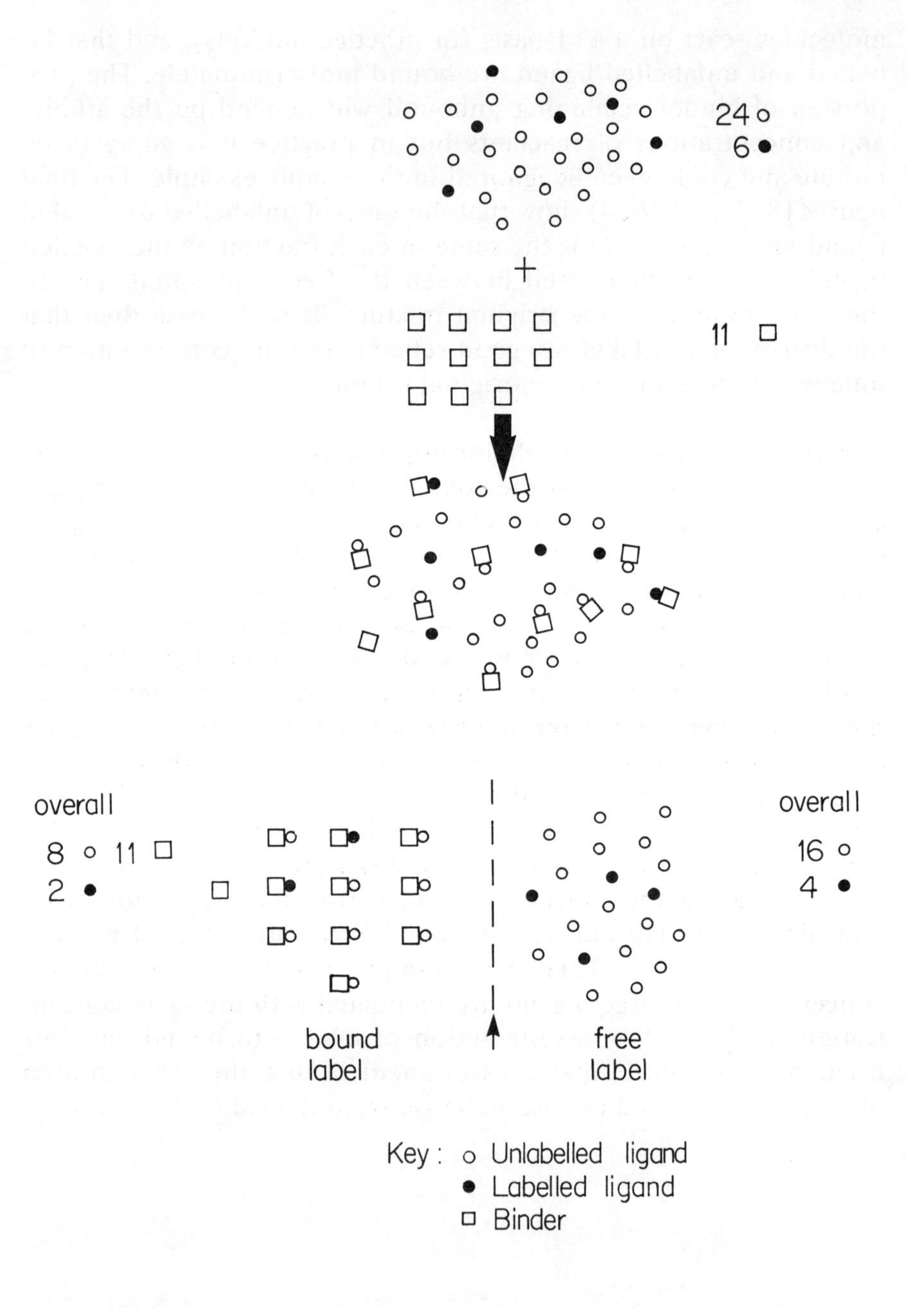


**Fig. 5.3c.** *Schematic illustration of the distribution of label in bound and free fractions of a labelled ligand assay*

(i) Effect of variation in the concentration of unlabelled ligand on the distribution of labelled ligand.

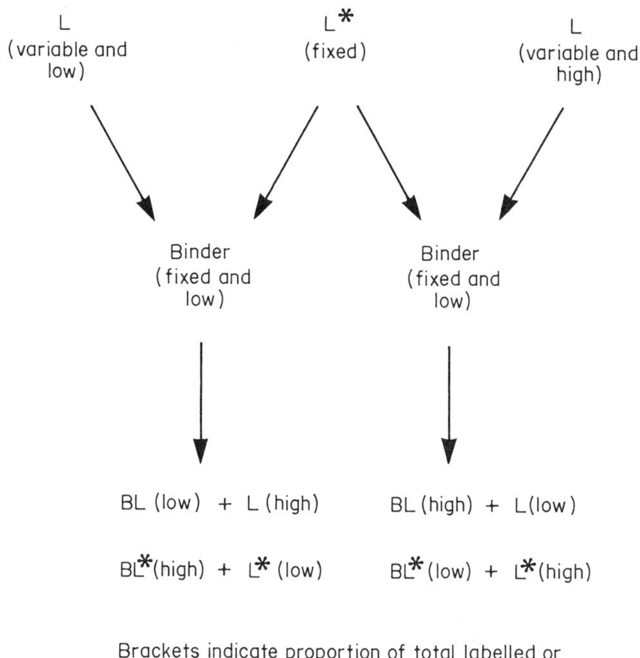


Brackets indicate proportion of total labelled or unlabelled ligand in that form.

(ii) Calibration curve

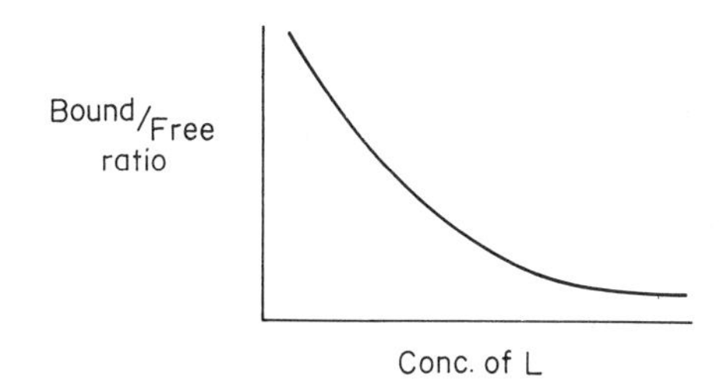


**Fig. 5.3d.** *The principles involved in the production of a calibration curve for a labelled ligand assay*

This type of assay requires:

- a suitable specific binder and a competition for it,
- a label which can be measured accurately, at low concentrations, and in either bound or free forms, or indeed in both, and
- a procedure for separating the bound and free forms.

### 5.3.1. Competition

Ligand assay methods of this type inevitably involve competition between labelled and unlabelled ligand molecules. These 'competitive' techniques have also been called saturation analysis techniques because the binder is virtually completely utilised due to its low limiting concentration. The optimal concentration of the binding reagent and the maximal performance are different for competitive assays (whether in labelled ligand assays or in the relatively few competitive labelled antibody assays) compared with assays described later in which there is no element of competition. Low limiting concentrations of binder are optimal for competitive methods whereas high concentrations are favoured in non-competitive methods.

To illustrate this point in the case of labelled ligand assays, consider an extreme situation in which the concentration of binder is much greater than that of the total ligand. It would then be the case that at equilibrium virtually all the ligand molecules, labelled and unlabelled, would be bound and any likely variation in the concentration of unlabelled ligand in the sample or standard would make no difference to the amount of labelled ligand bound (all of it) or free (none of it), Fig. 5.3e. Labelled ligand assays therefore, being competitive, are not favoured by excessive concentrations of binder relative to that of the ligand. The higher the concentration of binder the less sensitive the assay becomes.

The amount of binder has to be sufficiently small to leave an appreciable proportion, in general about 50% of the labelled molecules, free when the analyte concentration is zero. For this reason competitive methods are sometimes called limit binding reagent methods.

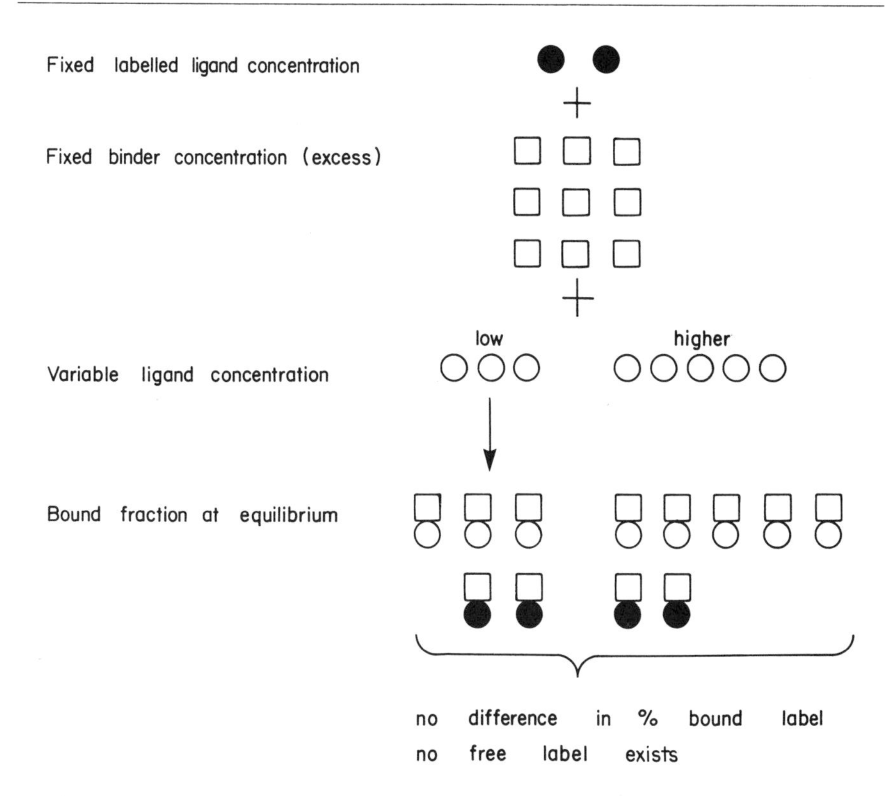


**Fig. 5.3e.** *The influence of binder concentration on a labelled ligand assay*

## 5.3.2. Purity of reagents

In biochemistry, labelled ligand assays have been longer established and are more widely used than the labelled antibody assays discussed later, though many analytes can be estimated by either method. The reason for this is historical. The tracer (labelled species) must be pure (ie ideally the only labelled material) whichever approach is used, otherwise a high background signal due to the other labelled components will result and binding reactions, other than the one required, could be included in the measurement. It is difficult to prepare antibody of sufficient purity for labelling from antiserum, so the general development of labelled antibody assays has been delayed until monoclonal antibodies became available relatively recently.

Antibodies, transport proteins, and cell receptors have all been used as binding reagents in labelled ligand assays, although undoubtedly antibodies are the most important of these. In labelled ligand assays, at least in theory, the binder does not have to be pure since only the reaction involving the labelled ligand will be measured, other independent binding reactions occurring at the same time being ignored by the measuring system. Unpurified antiserum can therefore usually be used as the source of antibody.

While labelled antibody assays are claimed to be more sensitive, labelled ligand assays are already extremely sensitive, certainly sensitive enough for many clinical analyses, and it is therefore unlikely that many well established labelled ligand assays will be quickly displaced.

**SAQ 5.3a**

Simple labelled ligand binding assays require three components:

- a sample for analysis or a standard ligand solution,
- labelled ligand,
- the appropriate binding agent.

(*i*) Which of the above is/are present at a fixed concentration?

(*ii*) Which need(s) to be present at low concentration?

(*iii*) Using the symbols L, L* and B, what are the four main components of the equilibrium mixture?

(*iv*) As the concentration of L rises, will the bound:free ratio of label rise or fall?

**SAQ 5.3a**

**SAQ 5.3b**

Indicate, with reasons, whether each of the following is true or false.

(*i*) In competitive binding assays the reaction is normally allowed to go to equilibrium.

(*ii*) In a simple system using a radioactive label it is necessary to separate the bound and free forms of the label in order to carry out the measurement. You need to consider (and perhaps therefore revise) the basic properties of radioisotopes and the principles of their measurement.

(*iii*) In binding assays the concentration of analyte in the samples can be determined by calculation from the known affinities of the binding agent by applying the Law of Mass Action.

**SAQ 5.3b**

**SAQ 5.3c**

By way of revision of material discussed earlier in this Unit list three reasons why antibodies are particularly useful as binding reagents.

## 5.4. DIVERSITY OF PROCEDURES

Labelled binding methods can be performed in a large number of ways. We have already seen that at least sometimes there can be a choice of binder (antibody, transport protein, or receptor). In this Section we introduce other possible variants, in particular the choice of label used, the reaction end point, whether or not a separation procedure is required, and the choice of component to be labelled.

### 5.4.1. Choice of label

The labels which were employed in most of the early methods were radioisotopic, but some years ago non-isotopic labels were introduced and rapidly gained in popularity. Choice of label is influenced by many factors and the label used may in turn determine the sensitivity of the method and various aspects of the experimental procedure. These aspects are discussed more fully in Part 6.

### 5.4.2. Equilibrium or disequilibrium procedures

In most methods, the binding reaction is carried to equilibrium, but because these methods are all comparative, with test samples being analysed alongside standards under the same conditions, it is not mandatory in principle to go to equilibrium. Procedures which stop short of equilibrium (disequilibrium methods) have been adopted for some assays in order to increase sensitivity. However to keep this account simple, these methods are not described further since at present they are of less importance than the equilibrium methods.

### 5.4.3. Separation or non-separation procedures

When the reaction is stopped, either before or at equilibrium, the tracer will be distributed between two states, free and bound in the complex, as discussed earlier and illustrated in Fig. 5.3b. It is then necessary to discriminate between the label attached to free and bound tracer for the final measurement. This can be achieved in one of two ways, either by physical separation of the fractions or without separation.

Originally all methods involved separation of the fractions before measurement because these early methods all involved radioisotope labels, which would of course emit their radiation equally well from both the free and bound state. Discrimination of the two states required therefore their separation by some physical or chemical means and such methods are frequently referred to as heterogeneous methods. The more recent introduction of a range of alternative labels (Section 6.2) has provided some situations in which the signal is substantially modified by the binding reaction, and when this is so, its measurement in one fraction might be possible without separation. Furthermore some methods are now available which are non-competitive and also do not require a distinct separation stage. Non-separation (homogeneous) methods are naturally quicker, potentially less error-prone, and generally more popular.

It is important to note that by whatever means discrimination between free and bound label is achieved, it is only necessary to measure the label in one of the fractions since the total amount of label is fixed.

### 5.4.4. Labelled ligand or labelled antibody

In binding methods a tracer label is used to demonstrate or quantitate a ligand when it reacts with a fixed amount of binder (the reagent). Tracer can be introduced as a fixed amount of labelled ligand or of labelled binder, but since in practice antibody is the only binding reagent which is labelled in current methods, procedures which use labelled binder can be referred to as labelled antibody methods. In current terminology labelled ligand methods are usually referred to as immunoassays, whereas labelled antibody methods are known as immunometric assays. Unfortunately the former term is also used in a very general sense to mean almost any technique involving a reaction with an immunological basis.

The two approaches differ in more than the nature of the labelled species itself since the choice of species influences the optimal conditions which need to be used for the actual technical procedure, and also the maximal performance of the method.

The importance of these labelled antibody methods warrants some further discussion.

## 5.5. ASSAYS WITH LABELLED ANTIBODIES

In the most straightforward design possible for a ligand assay using a labelled antibody, each tube would contain:

- the sample for analysis (or standard) containing a variable amount of ligand (L), and

- a fixed and relatively large amount of labelled antibody, B*.

This basic labelled antibody (immunometric) assay is summarised in Fig. 5.5a and in order to simplify this diagram the antibody is shown as univalent and any very small amount of ligand which remains free is ignored.

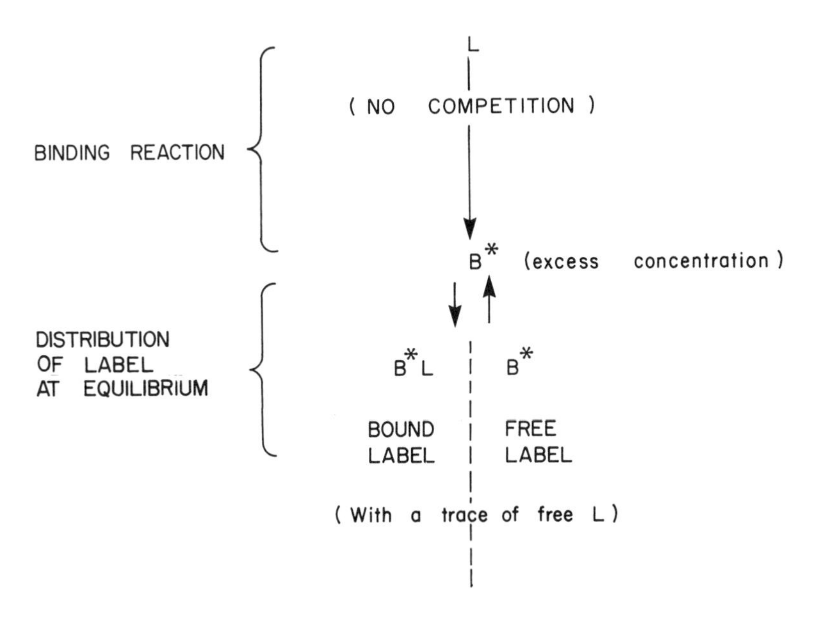


**Fig. 5.5a.** *Schematic diagram of an assay using labelled antibody. L is ligand, B* labelled antibody, B*L bound label*

Such a procedure would not depend on competition between labelled and unlabelled molecules of the same type for the binding agent, since in practice all of the ligand is taken up by the binder. The label is not distributed between bound and free fractions as before, but is all in the form of bound labelled binder. In this situation (in contrast to labelled ligand assays), a small change in initial ligand (analyte) will directly increase the amount of complex formed and the amount of label bound (Fig. 5.5b). In this non-competitive procedure, there is not the constraint on the relative amount of binder, that was found in the competitive labelled ligand assays. In fact such non-competitive procedures are optimised with a concentration of binder which is high relative to ligand, so that all the ligand is bound at equilibrium. Such non-competitive labelled antibody assays are therefore examples of excess reagent methods.

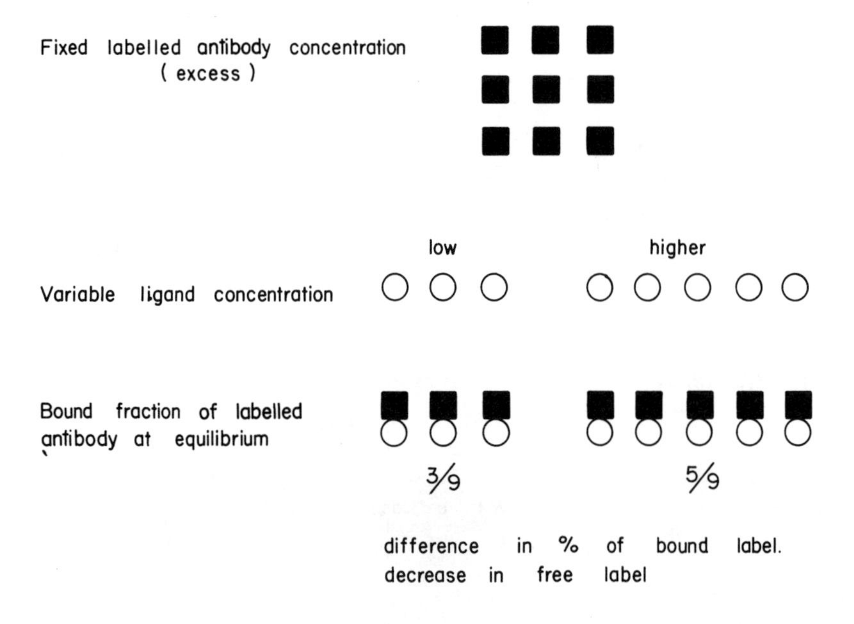


**Fig. 5.5b.** *The influence of binder concentration on a labelled antibody assay*

A comparison of Figs. 5.3e and 5.5b shows the situation with an excess of binder, at two concentrations of analyte. Under these conditions the labelled ligand assay is insensitive to analyte concentration but the labelled antibody assay is sensitive. In the presence of insufficient binder the reverse would hold.

Like labelled ligand assays, all labelled antibody assays are comparative, a range of standards always being included in each analysis batch to produce a calibration curve for the measurement of the sample analyte by interpolation.

All the tubes, within an assay, ie samples and standards are treated identically. In addition to having identical amounts of antibody and label, all receive identical treatments with respect to final volumes per tube, incubation time and temperature and separation methods, where applicable.

### 5.5.1. Two site assays

The labelled antibody assays used in biochemical analysis are frequently more involved in design than the simple procedure shown in Fig. 5.5a. An example, the two-site immunometric assay, is illustrated in Fig. 5.5c. This non-competitive labelled antibody assay uses two antibodies, $Ab^x$, and $Ab^y$, specific to different sites, x and y, on the ligand antigen $Ag^{x.y}$. In this case, one antibody ($Ab^y$) is bound to a solid phase and effects the separation of $Ag^{x.y}$ from others (eg $Ag^z$). The other is labelled, ($Ab^x-*$), and on addition in excess will bind to all the $Ag^{x.y}$ molecules thus labelling them for quantitative measurement.

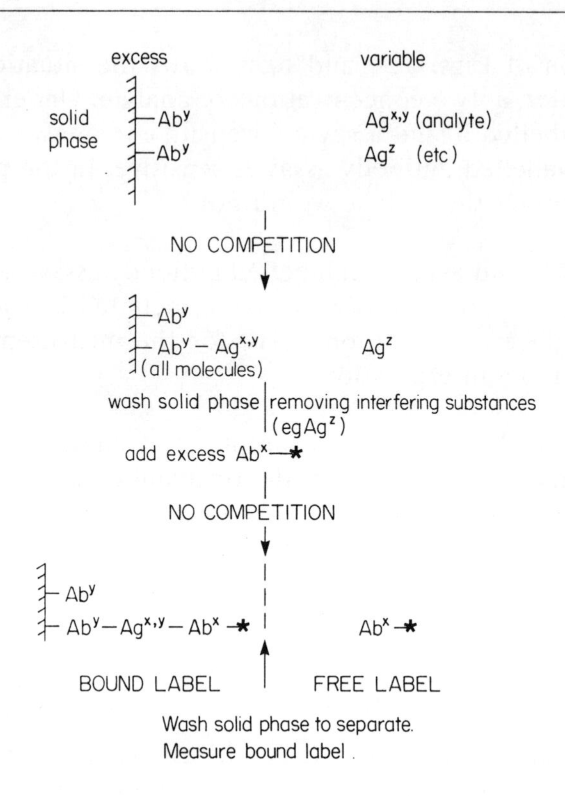


**Fig. 5.5c.** *Non-competitive labelled antibody assay (two-site immunometric assay)*

This approach is sometimes referred to as a sandwich technique, since the target analyte is being sandwiched between two antibody molecules in the final complex. One of the several advantages of two-site sandwich assays is that the first binding reaction, yielding

$$\text{Solid Phase-Ab}^{y}\text{-Ag}^{x,y}$$

in effect acts as an initial extraction of all of the analyte from the sample as well as serving as the basis of the subsequent separation of the free and bound label. Another advantage is the high specificity that can be obtained through the use of two distinct antibodies to the same target analyte antigen molecule.

The so-called gonadotrophin hormones provide a particularly good illustration of the advantage of this development since several members of this group show considerable structural similarity leading to potential difficulty in their separate measurement. The thyroid stimulating hormone (TSH), follicle stimulating hormone (FSH), luteinising hormone (LH) and human chorionic gonadotrophin (HCG) all possess identical $\alpha$-protein sub-units and only vary in their $\beta$-units. Methods have been developed, for example for TSH, where the $\alpha$-unit can be bound to an anti-$\alpha$-unit antibody attached to a solid phase thus obtaining a separation of these hormones from other sample constituents, followed by a specific measurement of TSH using a labelled (*) anti-TSH $\beta$-unit antibody, thus:

$$\text{Solid phase} - \text{anti-}(\alpha\text{-unit}) - \text{TSH} - \text{anti-}(\beta\text{-unit})^*$$

Though in principle the procedure could be carried out using purified polyclonal antibodies, one attached to a solid support and the other carrying the label, these assays are usually performed with two monoclonal antibodies specific to different, well separated sites on the antigen as shown in Fig. 5.5c. Obviously this technique is not directly applicable to haptens.

A further development of the sandwich principle is to use two antibodies ($Ab^x$ and $Ab^y$) as before, in order to obtain the advantages mentioned above, but to introduce the label with a third antibody ($Ab^z$—*). If this antibody is a general anti-IgG antibody then it can be used in a wide range of different analyses since it does not have to be specific for the target antigen. The final end product is:

$$\text{Solid phase -}Ab^y\text{-}Ag^{x,y}\text{-}Ab^x\text{-}Ab^z - *.$$

### 5.5.2. Competitive procedures using labelled antibody

While labelled antibody assays are, in the main, non-competitive and require excess reagent, it is possible to devise labelled antibody assays for the estimation of ligand in which there is competition.

For example, ligand in the sample for analysis could be made to compete with a fixed amount of ligand, which is also unlabelled but is bound to a solid phase, for a limited amount of a labelled free antibody. The label will distribute itself between the bound and free fractions with increasing quantities of it in the latter as the concentration of free antigen rises (Fig. 5.5d). The label in one fraction, is measured after separation. In this case the optimal concentration of labelled antibody is low relative to ligand, ie this is a limited reagent method.

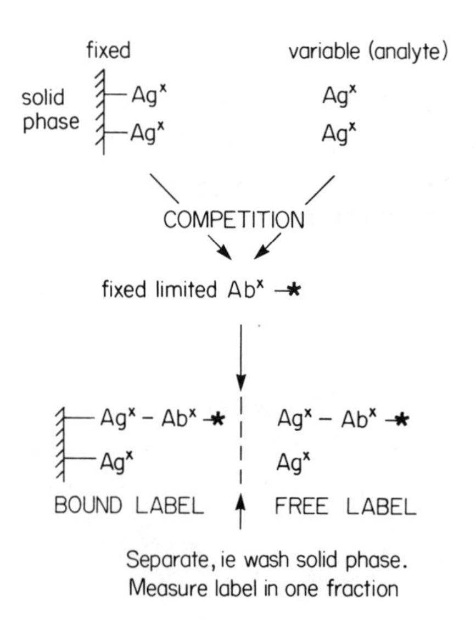


**Fig. 5.5d.** *Competitive labelled antibody assay*

Competitive labelled antibody procedures of this or other designs are increasingly used in various laboratories. They have the advantage over sandwich type assays of requiring only one type of antibody, and using low concentrations of it will further decrease the cost of the assay. Additionally the requirement for only one wash step simplifies the assay and reduces the time and error. However, non-competitive immunometric assays should be capable of greater sensitivity and specificity. More involved designs, both competitive and non-competitive, using labelled antibody are described in Part 6.

### 5.5.3. Advantages of labelled antibody assays

The use of polyclonal antibody (from antiserum) as a binding reagent in labelled antibody procedures is difficult as the tracer must be pure and the purification of antiserum is technically difficult. Furthermore large amounts of antiserum are required both because of the losses in purification and because of the excess needed for many assay procedures.

The availability of large amounts of pure monoclonal antibodies has reduced these problems, and the situation now is that immunometric assays have a number of potential advantages over labelled ligand procedures which are making them increasingly popular.

The assays frequently have high precision due to many factors. For the majority of procedures the number of pipetting steps is low compared with competitive immunoassay (IA) methods and furthermore since the antibodies are present in excess the immunometric assays (IMA) are less susceptible to pipetting errors. The excess antibody also makes the methods less susceptible to matrix effects.

While the dose/response curve for an IMA is usually steep giving useful sensitivity, the effective assay range is often linear over a much longer span of concentrations than many competitive IA methods with only limited antibody concentrations. This can be important since the concentration of some analytes can vary over several orders of magnitude, ferritin ranging from 4–1000 $\mu g\ dm^{-3}$ of serum for example.

The use of excess antibody, so that all antigen molecules take part in the reaction, the facility to label the large antibodies with a high density of measurable label, the availability at the present time of antibodies with high specificity and affinity, and the high precision referred to above, all contribute to a usefully low detection limit. These limits are typically at least one order of magnitude lower than for standard IA methods.

At a more practical level the use of labelled antibodies is advantageous because:

- they are more easily labelled than many smaller ligands which may have a limited number of target groups for labelling and which may suffer more significant chemical changes on labelling,
- higher stability of the labelled species in physiological fluids and on storage,
- faster reactions due to them being driven by the excess of antibody.

Whether these advantages are sufficient for the methods to replace entirely the established labelled ligand assays is as yet difficult to say.

**SAQ 5.5a**

Why is it necessary to carry out a separation of bound and free forms of the tracer before measurement when radioisotopes are used as the label?

**SAQ 5.5b**

Consider a non-competitive two-site labelled antibody assay.

(*i*) Which component is bound to the solid phase?

(*ii*) What proportion of sample antigen is bound to the first antibody?

(*iii*) How is the separation of sample antigen from the others achieved?

(*iv*) Which materials must be present in excess?

(*v*) Illustrate why this technique is sometimes referred to as a sandwich technique.

**SAQ 5.5c**

(*i*) Complete the following sentence with either the term competitive or non-competitive.

All labelled ligand assays for the estimation of ligand are ________ whereas most labelled antibody methods are ________

(*ii*) Complete the following sentence with either the term, limited or excess.

Competitive assays perform optimally with ________ reagent, whereas non-competitive assays are optimised with ________ reagent.

## 5.6. APPLICATIONS OF LABELLED BINDING ASSAYS

The advantages of labelled binding assays over other assay techniques may be exemplified by the very wide range of applications to which the former are commonly put. While they have found very widespread use in general biological situations, in a clinical context the applications include:

- the assay of proteins, including virtually all human peptide hormones, antibodies themselves, blood congulation factors, and antigens characteristic of hepatitis B and cancerous cells;

- the assay of drugs, antibiotics and other xenobiotics including those present at toxic levels;

- the assay of a very diverse range of complex organic materials of clinical relevance, eg progesterone, testosterone (and other steroid hormones), cyclic AMP and cyclic GMP (intra-cellular regulation factors), folic acid and $B_{12}$ (vitamins), and 'intrinsic factor' (which is involved in intestinal vitamin $B_{12}$ absorption).

The procedures are applicable to fluids such as plasma, urine, saliva, amniotic fluid and to cell and organ extracts. In a clinical situation the samples are usually of complex human extracts, containing analytes at very low concentrations, in the presence of many other related and unrelated molecules, and furthermore large numbers of samples may need to be analysed in a relatively short time.

Consider for example the drug digoxin which is given to patients suffering from myocardial infarction (the most usual type of heart attack). The drug is very potent at low levels in the body but may cause serious side-effects if blood levels exceed a certain concentration. It is imperative therefore that the clinician gives a dose of drug that will ensure that the correct concentration appears in the patient's blood, bearing in mind his body size and the effect of his age on the body's tendency to metabolise and excrete the drug. Regular blood assays are therefore required.

Immunological assays are the method of choice since the volumes of plasma or serum samples taken for analysis are likely to be small (say 0.1 $cm^3$, 100 $\mu l$), and these methods do not require extraction of the drug from the sample and can generate a result quite quickly. Hence many such analyses can be performed in a short time and relatively cheaply which is just as well considering the all too common occurrence of this condition in Western populations.

For digoxin the optimum circulating levels are about 1 ng $cm^{-3}$ (1 mg $dm^{-3}$) and therefore 100 $\mu l$ aliquots of serum or plasma will only contain about 100 pg of the drug. Many other analytes such as the protein hormones also circulate at concentrations of this order and to detect these materials by most other methods, eg chromatographic procedures, many $cm^3$ of blood would be required to allow concentration by extraction prior to the analysis. Even where such extractions are possible they are time-consuming and generate additional sources of error.

**SAQ 5.6a**

You are required to assay the anti-epileptic drugs phenytoin, valproic acid, and carbamazepine in the blood of epileptic infants. These drugs are commonly co-administered and the range of their serum levels is 10–100 $\mu g\ cm^{-3}$ (10–100 $mg\ dm^{-3}$). Based upon the information given previously and your general knowledge of the design of analytical experiments, explain, with reasons, which assay system you would choose from those listed below.

(*i*) Individual immunoassays for each drug requiring 10 $\mu l$ of sample per assay with each assay being automated and taking 5 minutes to perform.

(*ii*) One single HPLC assay which co-determines each drug in one run. A sample volume of 50 $\mu l$ and 30 minutes assay time are required.

(*iii*) Chemical derivatisation procedures leading to spectrophotometric analyses with 10 $cm^3$ original sample and 45 minutes of time required for extraction, reaction and measurement.

**SAQ 5.6a**

**Summary**

Methods involving changes secondary to the binding reaction, especially those resulting in the formation of aggregates, the agglutination of cells, and the rupture of cells by complement fixation are only briefly considered.

The requirements for satisfactory assays using labelled ligands, the importance of the principle of competition and the distribution of label into free and bound forms, and some fundamental variables in the basic systems are considered in more detail.

The advantages of using labelled antibodies in immunological assays, and some of the special variants of these techniques are described.

**Objectives**

You should now be able to:

- describe the types of method which involve changes secondary to binding reactions;
- list the end-results of these assays and hence the methods used in their measurement;
- state and discuss the basic requirements for an assay using a labelled ligand;
- explain the importance of competition, reagent purity, and selection of the appropriate binder concentration in these assays;
- discuss the nature of important variables such as choice of label, choice of component to be labelled, the necessity for separation of free and bound forms of labels, and whether or not the system reaches an equilibrium state;
- explain the relationship between quantity of label in the bound and free forms and quantity of analyte in a test sample or standard mixture;
- discuss the necessity for, and merits of, assays using labelled antibodies, especially the two-site immunometric systems;
- understand that in general such systems are non-competitive but that competitive variants are available;
- appreciate the general importance of labelled binding methods in clinical analysis.

# 6. Technical Details of Some Labelled Binding Assays

## Overview

This Part of the Unit will consider some important practical aspects of the immunoassay techniques discussed in earlier Parts, in particular the relative merits of the labels available and some of the procedures used to study and evaluate a newly developed method.

## Introduction

This Part gives some technical details of a selection of the great variety of labelled binding assay procedures which can be applied to the analysis of ligands. In the earlier Parts (3 and 4) you will have learned of the specificity of the binding interaction between an antibody and its antigen or hapten, and between a biological receptor or transport protein and its ligand. Then in Part 5 it was shown how we can exploit these interactions to develop assays of great sensitivity and versatility which use a labelled reactant to monitor these binding processes. Here we will focus mainly on assays using labelled ligands and give some attention to the variety of labels and measurement techniques used in these procedures and to some of the practical aspects of developing, carrying out and monitoring these procedures.

## 6.1. SEPARATION ASSAYS WITH LABELLED LIGANDS

Since these were historically so important and are still of great significance they represent a useful starting point and warrant a specific example by way of illustration. In these systems a fixed quantity of labelled ligand is incubated with the analyte (which is unlabelled ligand), and a fixed but insufficient amount of binder. After reaction, the bound and free label are physically separated before one or other is measured. As an example of this approach let us take the radioimmunoassay of human placental lactogen in serum.

### 6.1.1. The radioimmunoassay of human placental lactogen (HPL)

HPL is a proteinaceous hormone secreted by the placenta, the tissue connecting a pregnant mother to her unborn child. The embryo child receives all its nutrients and oxygen via the placenta and its well-being is absolutely dependent on adequate placental function. HPL can be used as one of the indicators of placental function during pregnancy since the placenta is the sole source of the hormone and the blood level directly reflects the total mass of secreting, ie functional, tissue.

Like most other proteins HPL can be determined by radioimmunoassay since it is antigenic and antibody can be produced by immunising an animal of a different species as described earlier. It is useful that successful methods have been developed using unpurified antiserum and that the hormone itself can be iodinated with $^{125}I$, an emitter of low energy $\gamma$-radiation, through its tyrosine residues.

In this particular method separation is achieved by ethanol precipitation of the bound label, ie the antibody-HPL complex, and the amount of radioactivity in this precipitated complex can be measured. It is not the purposes of this text to give detailed methodological instructions, but the overall procedure is summarised in Fig. 6.1a.

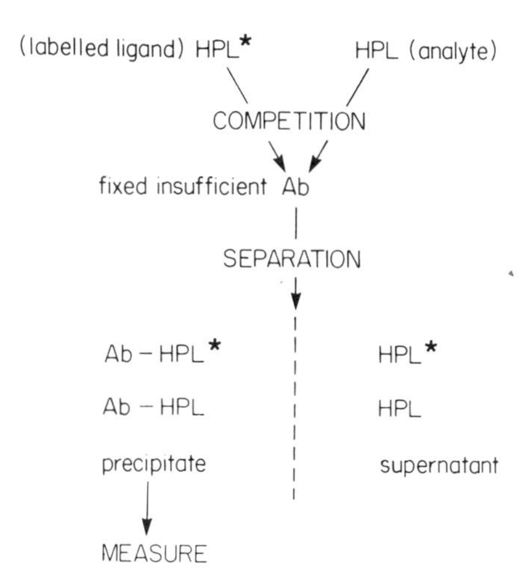


**Fig. 6.1a.** *An illustration of the competitive labelled ligand radioimmunoassay for human placental lactogen*

**SAQ 6.1a**

Some typical results from a competitive labelled ligand radioimmunassay are tabulated.

| Standard $\mu g\ cm^{-3}$ | Radioactive counts in the precipitate, duplicate results, bound fraction (in disintegrations $s^{-1}$) | |
|---|---|---|
| 1 | 62 011 | 61 980 |
| 3 | 46 082 | 45 927 |
| 6 | 30 615 | 30 950 |
| 8 | 24 959 | 24 659 |
| 10 | 20 981 | 21 596 |
| Test sample | 36 542 | 37 218 |

Plot the results of the standards on the graph paper below using the vertical axis for the radioactive counts. Draw the best curve you can, using a 'flexicurve' if one is available. Determine the concentration in the test serum by interpolation from the graph.

**SAQ 6.1a**

### 6.1.2. A general procedure

A general procedure for a separation labelled ligand assay would therefore be along the following lines.

- Add aliquots of standard solutions (covering the range of concentrations expected to be present in the test samples, bearing in mind the possible abnormal state of the latter in a clinical situation) and of the test samples themselves, to separate assay tubes.

- Add labelled ligand to each of the tubes in the batch.

- Add binding reagent and then vortex mix the tubes to obtain a thorough mixing of reagents.

- Leave the tubes to incubate for a predetermined period of time at a temperature appropriate for the assay.

- Add a reagent to each tube or carry out some other procedure which effects a separation of the bound labelled ligand from the unbound label.

- Determine the radioactivity or measure some other signal of one of the fractions (bound or free).

- Plot a standard curve of the measured label against the concentrations of standard analyte. Determine the concentration of analyte in the samples from the graph.

The data from experiments such as these can be plotted in a variety of ways, eg with % bound label or perhaps the ratio of free/bound label as the y axis. An awkward problem is that the standard graph is generally curved when the $x$ and $y$ axes are on a linear scale. This present the obvious problem of drawing the best line through data which are likely to be less than perfect; hence the recommendation to use a flexicurve! A less obvious problem but one that is important in a laboratory which routinely uses any method, concerns the ease with which the essential regular recalibration can be carried out. If a plot is linear, providing the slope of the response is known recalibration can generally be done by the measurement of just one standard sample, which is plotted and a line drawn with the appropriate slope. Recalibration of a curve is much more awkward and generally needs the measurement of more samples, since it is likely to be the case that the algorithm for the plotting of the curve will be unknown.

Much effort has gone into finding ways of transforming data to enable them to be plotted in a linear fashion, and one widely used approach in this field is the logit plot. Here the signal is plotted in the form of logit $Y$, where logit $Y = \ln[Y/(1 - Y)]$ and $Y$ is the proportion of the label that is bound at a given analyte concentrations. Log $X$, where $X$ is the analyte concentration, is plotted on the abscissa ($x$ axis).

**SAQ 6.1b**

Replot the data provided in SAQ 6.1a, given that the total radioactivity was 75 000 disintegrations $s^{-1}$, using the logit $Y$/log $X$ plot described above. To help you in this, note that the mean radioactivity of the 1 $\mu g\ cm^{-3}$ standard is 61 996 and hence the proportion bound is 61 996/75 000 or 0.827. Hence determine the concentration of the test sample.

## SAQ 6.1b

## 6.2. LABELS FOR IMMUNOASSAYS

The tremendous value of immunoassays in biological and other investigations is perhaps partly illustrated by the enormous range of methods that has been developed using the principles outlined in the earlier parts of this Unit. Among these published methods one can see a very wide range of labels which have been used by different workers confirming the importance of these techniques right across the field of biological and related sciences. In principle any material which has certain properties could be used as a label and perhaps the most important of these properties are:

- the label should be cheap to buy, safe to use and be amenable to labelling procedures which are simple, rapid and not needing purification after labelling;

- the label needs to be covalently linked to the ligand and preferably via multiple sites to give a high specific activity;

- the labelled ligand must be stable;

- the label should have minimal effect on the properties of the ligand, so that labelled and unlabelled forms behave identically;

- the label should be detectable by a simple, sensitive, cheap, automatable method which is not influenced by sample or reagent variables and preferably not by environment variables either;

- the label should ideally have properties which enable one to differentiate it in the free and bound forms to avoid the need for separation stages.

The following list shows some of the labels which have been used in published methods:

radioisotopes,
fluorescent groups
enzymes,
coenzymes and prosthetic groups,
red blood cells, latex and other particles,
viruses,
metals,
free radicals,

As has been discussed earlier the most common labelled ligand procedures can be sub-divided into those in which the bound label is physically separated from the unbound (separation or heterogeneous assays), and those that do not require separation (non-separation or homogeneous assays). Since all the early procedures used radioisotope labels, which do not show a change of signal with change of physical or chemical state, a separation was obligatory. The catalyst for the introduction of the alternative labels shown in the list above was the desire to develop simpler, non-separation procedures. Obviously therefore the labels used in immunoassays have had considerable impact on the techniques and warrant discussion in some detail.

### 6.2.1. Radioisotopic labels

Perhaps because these materials were the first labels to be used to any significant extent, it is the case that more methods employ them than any other although some non-radioisotope labels have gained in popularity in recent years and may well supplant radioisotopes in many applications.

Radioisotopes have some significant advantages as labels, perhaps most noticeably the ease with which they can be detected at very low levels, the relative simplicity of the labelling techniques, and the minimal effect of the label on the properties of the ligand. However the cost of radioisotopes and their measuring systems is reasonably high, and rising rapidly, and some workers consider that there is a significant personal hazard in their use. In addition the monitoring and disposal of radioactive materials represents an added inconvenience. Certainly there is extensive legislation involved in their use.

Radioisotopes suffer from the phenomenon of isotopic decay which results in the gradual disappearance of radioactive atoms and a reduction in count rate. Two common isotopes, $^{125}I$ and $^{131}I$ have half-lives of 60 days and 8 days respectively and as a consequence, no matter how intrinsically stable a labelled component may be it will need replacing at regular intervals. This adds considerably to the cost of the assay and the problems of assessing and maintaining the quality of the assay.

Two major radionuclides have found widespread use in labelled ligand assays, iodine-125 ($^{125}I$) and tritium ($^{3}H$). Currently for radioreceptor assays and immunoassays of small haptens, tritium is generally used, whereas for radioimmunoassays of large molecules, most commercial kits and in-house methods employ $^{125}I$. The latter is the isotope most commonly used to label proteins since it can be attached to the aromatic rings of tyrosine residues or to histidine residues in proteins, following oxidation by chloramine T or preferably by less destructive methods using lactoperoxidase enzymes. Even less interference with the target molecule can be obtained by coupling an already labelled molecule to the target (conjugation labelling).

Procedures for the labelling of an antigen are given below. If you have an interest in chemistry you might care to work out the chemical reactions involved in these processes. Reference to some of the texts (eg Edwards, 1985) cited in the Bibliography may be of help.

| | Chloramine T method | | Solid-phase lactoperoxidase method |
|---|---|---|---|
| ADD | 10 $\mu$g antigen in phosphate buffer pH 7.4 to 1 mCi $^{125}$I as sodium iodide | | |
| ADD | 10 $\mu$g chloramine T (10 $\mu$g) | or | 10$\mu$l lactoperoxidase (10-20 ng) + 5 $\mu$l hydrogen peroxide (0.5 nmol) |
| WASH | 30 s | | 10 min |
| ADD | 0 | | 5 $\mu$l hydrogen peroxide (0.5 nmol) |
| ADD | 10 $\mu$l sodium metabisulphite (10 $\mu$g) | | 0 |
| INCUBATE | 0 | | 20 min |
| ADD | 0 | | 10 $\mu$l sodium azide (100 $\mu$g) |
| ADD | 100 $\mu$l potassium iodide in 1% albumin (1 mg) | | |

Assays for proteins low in tyrosine residues, or for non-protein materials such as the steroid hormones, require $^{3}$H labelling or the covalent linking of a tyrosine- or histidine-containing moiety or moieties to the hapten or antigen. This moiety may be iodinated with $^{125}$I prior to binding or the derivative subsequently iodinated. Alternatively in some specialised systems isotopes of cobalt, iron and even selenium may be used. In spite of its longer half-life which would help reduce the cost of the assays by extending the usable lifetime of the reagent, tritium has found few applications due to the costs involved in its detection (liquid scintillation as opposed to gamma-ray spectrometry). Of course this point is invalid if an institution requires a scintillation counter for other purposes. Its lower specific activity however can result in a lower assay detection limit and it should be noted in this context that one atom of $^{125}I$ gives six times as much detectable radiation as one atom of $^{3}H$.

| Radio-nuclide | Symbol | Half-life | Emission | Specific Activity Ci mg atom$^{-1}$ (Bq mg atom$^{-1}$) |
|---|---|---|---|---|
| Tritium | $^{3}H$ | 12.35y | $\beta$ | 29($1.07 \times 10^{12}$) |
| Iodine-125 | $^{125}I$ | 60d | $\gamma$ and X | 2 176($0.78 \times 10^{15}$) |

**Fig. 6.2a.** *Comparison of some properties of the two most commonly used radioisotopes*

**SAQ 6.2a**

For an assay using a labelled ligand explain why $^{125}I$ will produce a method with lower detection limits than one with $^{3}H$.

### 6.2.2. Non-isotopic Photon-emission labels

In the case of protein radioimmunoassays the $^{125}I$ atom is introduced into a tyrosine or histidine residue in the molecule. As an alternative it would be useful if this could be replaced by a material detectable by some spectroscopic method, eg fluorimetry. We could then carry out the general procedure described previously but using this label, and obtain a signal by measuring the label in one of the fractions using a fluorimeter. The standard curve would be a plot of fluorescent intensity against concentration, and sample concentration would be reflected in changes in fluorescence also. In these basic systems, materials such as fluorescein, rhodamine and umbelliferone are commonly used as the fluorescent label, and competitive and non-competitive methods, and antigen and antibody assays, are all available. As you will see in the following text there is a wide range of methods available using these labels and a significant advantage of them all is that the reagents and reactions are generally regarded as safe and do not need special licensing or handling procedures.

While the lower detection limit for methods involving radioisotope labels is about $10^{-12}$ mol cm$^{-3}$, that for simple methods using fluorescein as a fluorescent label is only about $10^{-9}$ mol dm$^{-3}$, and perhaps the main reasons for this poorer quality are the scattering of incident and fluorescent light rays, and the quenching of fluorescence, due to the sample and instrument components. Background fluorescence giving a significant signal:noise ratio is another problem that is particularly troublesome in biological specimens due to their high content of proteins and other fluorescing materials.

One advantage of fluorescence as an immunoassay label however is that the fluorescent principle can be used in a number of different ways to produce a measurable signal in the assay method, some of which produce excellent specificity, detection limits and/or sensitivity and help minimise the consequences of the above problems.

Some of these alternatives are considered in Fig. 6.2b. In all cases competition occurs between sample antigen ($Ag_{(S)}$) and fluorescently labelled antigen (Ag-F) for the limited antibody (Ab).

Separation of bound and free label is effected and the fluorescence of one fraction (usually the free form) is measured. An increase in the concentration of Ag in the competion reaction mixture results in a decrease in Ag-F binding to the Ab. The percentage Ag-F(free) will therefore increase.

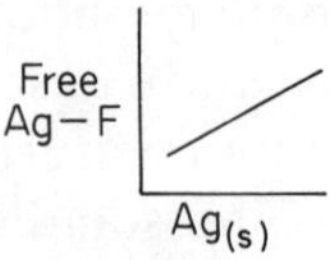


Basic System : $\text{Ab} + \text{Ag-F} + \text{Ag}_{(s)} \longrightarrow \text{Ab-Ag-F} + \text{Ag-F} + \text{Ab-Ag}_{(s)} + \text{Ag}_{(s)}$

In all cases % Ag− F (free) rises with increasing $\text{Ag}_{(s)}$

* indicates presence and extent of fluorescence

**Fig. 6.2b.** *Alternative applications of the use of fluorescent labels in immunoassays*

(*i*) Enhancement fluorescence

$$\text{Ab} + \text{Ag-F}^{*} \longrightarrow \text{Ab-Ag-F}^{**} + \text{Ag-F}^{*}$$

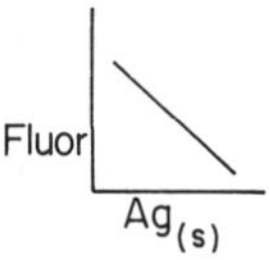


Here the fluorescence intensity increases (represented as $F^* \rightarrow F^{**}$) on binding to an antibody, hence the prevention of such binding by sample antigen ($Ag_{(s)}$) results in a decrease in fluorescence. While an assay method for thyroxine exists in which a four-fold rise occurs, such methods are relatively rare, but interestingly thyroxine is one of the few materials that can be analysed by an enhancement enzyme immunoassay (see later).

(*ii*) Direct quenching fluorescence

$$\text{Ab} + \text{Ag-F}^{*} \longrightarrow \text{Ab-Ag-F} + \text{Ag-F}^{*}$$

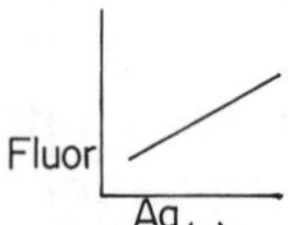

The fluorescence decreases (quenches) on binding to the antibody. These methods are much more common but in some cases the change in fluorescent signal is quite small, especially when the ligand is relatively large and effectively shelters the fluorescing label.

(*iii*) Indirect quenching fluorescence

$$\begin{array}{ccc} Ab + Ag - F^{*} \longrightarrow & Ab - Ag - F^{*} & + \; Ag - F^{*} \\ & Ab_F \rightarrow \downarrow & \rightarrow \downarrow \\ & Ab - Ag - F^{*} & Ag - F - Ab_F \end{array}$$

Fluor vs $Ag_{(s)}$

In these methods an antibody to the fluorescent label itself ($Ab_F$) is produced, and after the equilibrium step this antibody is added to the system. Only the free, unbound fluorescing antigen molecules react with this antibody and are quenched due to the presence of the large antibody molecule. Those labelled antigen molecules attached to the first antibody are protected from the second antibody and are not therefore quenched. The extent of quenching is then a reflection of the proportion of free label.

(*iv*) Enzyme release fluorescence

$$\begin{array}{ccc} Ab + Ag - F \longrightarrow & Ab - Ag - F & + \; Ag - F \\ & \text{Enzyme} \rightarrow \downarrow & \rightarrow \downarrow \\ & Ab - Ag - F & + \; Ag + {}^{*}F \end{array}$$

Fluor vs $Ag_{(s)}$

This system uses labels which do not fluoresce when bound to antigens; the fluorescence appearing when the label is released by enzymic digestion. The binding of an antibody during the course of the immunoassay prevents the enzymic release and thus the extent of fluorescence following enzymic digestion is inversely proportional to the extent of antigen-antibody binding. This approach has been developed by the AMES Division of Miles Laboratories into a dry phase (film) assay for a range of analytes including the anti-asthmatic drug theophylline.

(*v*) Fluorescent excitation transfer

$$Ag-{}^{*}F_1 + Ab-{}^{*}F_2 \longrightarrow \underset{\lambda ex_1}{\overset{\lambda em_1}{F_1}}-Ag-Ab-\overset{\lambda em_2}{F_2} + Ag-\underset{\lambda ex_1}{\overset{\lambda em_1}{F_1}}$$

Fluor

$Ag_{(s)}$

The antigen and antibody molecules are labelled with different fluors, ($F_1$ and $F_2$) which are carefully chosen to have overlapping spectra so that internal quenching will occur. If the molecules are sufficiently close together, some of the emission energy of one fluor will be transferred to the other resulting in a reduction (quenching) of measurable radiation at the emission wavelength ($\lambda_{em_1}$) of the first fluor. Thus if the equilibrium mixture is illuminated with the excitation wavelength ($\lambda_{ex_1}$) of one of the fluors, the extent of quenching of the emission wavelength ($\lambda_{em_1}$) of this fluor is directly proportional to the amount of antigen-antibody binding.

Two-site sandwich assays using two antibodies labelled with different fluors have also been developed. The Syva Corporation have introduced such a method for the assay of digoxin (a digitalis derivative widely used in the treatment of heart conditions and one of the most frequently measured therapeutic drugs), that uses the pigment phycoerythrin isolated from red algae as the primary fluorescent label. It absorbs about 30 times more light than fluorescein and re-emits 80% of it. With a fluorescein derivative as the second fluor, the method has the useful characteristics of detection limits of $<0.5$ mg dm$^{-3}$ and a precision of $<6\%$ coefficient of variation.

Two site sandwich

$$---Ab_1-F_1 + Ag \longrightarrow --\underset{F_1}{Ab_1}-Ag \xrightarrow{Ab_2-F_2} ---\underset{F_1}{Ab_1}-Ag-\underset{F_2}{Ab_2}$$

(*vi*) Fluorescent polarisation assays

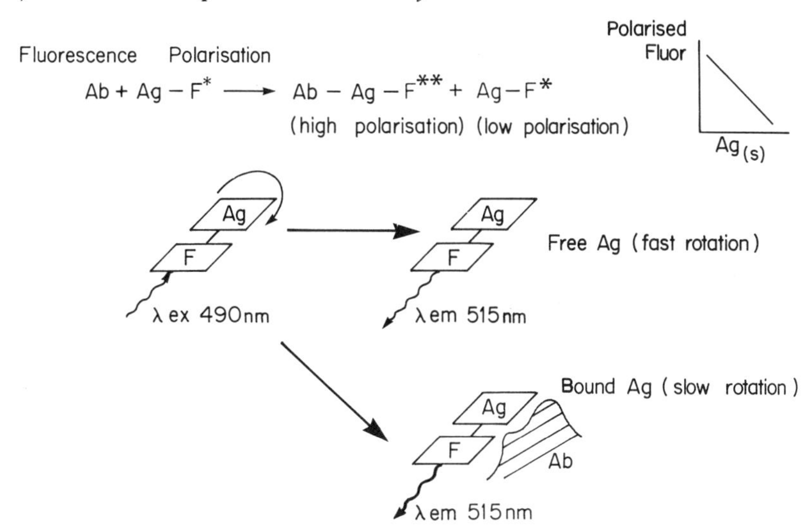


Let us take a relatively low molecular mass complex such as a phenobarbitone – fluorescein conjugate ($M_r$ about 500), as an example of the principle of this approach. If plane polarised light of the excitation wavelength for fluorescein (490 nm) is shone onto a solution of the conjugate, those molecules which lie in the same plane as the excitation beam will be excited. However the molecules will be rotating quite rapidly in the solution due to their small size, and in fact the time taken for them to rotate out of the plane of the incident radiation is less than the time spent by the electrons of the fluorescein moiety in the excited state, prior to their release of radiation at the emission wavelength of 515nm. The emitted radiation is likely therefore to be randomly orientated and not be in the plane of the excitation radiation; the signal is said to be depolarised. Compared with the phenobarbitone conjugate the antibody is relatively massive ($M_r$ about 125 000) and on binding the speed of rotation of the conjugate will decrease considerably. In fact the rotation time will become greater than the time spent by the electrons in the excited state and a much higher proportion will be emitted in the same plane as the incident radiation.

The general observation is therefore that the antibody – bound tracer will exhibit a higher degree of polarisation of fluorescence compared to that of the unbound tracer, and as a consequence we have a very simple method for determining the amount of ligand bound to the antibody. The procedure has the advantage of not requiring a separation stage since the binding event produces a substantial, measurable change in signal. Interestingly it is one of the few immunioassay techniques that is more suitable for the analysis of materials of low rather than high molecular mass. The reason for this is that the extent of polarisation of emitted light varies as the molecular size and amount of rotational motion, consequently large protein molecules exhibit a greater degree of polarisation than do small haptens. The binding of an antibody results in a larger relative change of polarisation for the lower molecular mass compounds.

Π By way of revision or a break perhaps you could look at the sequence for the separation assay described in Section 6.1.2. What step(s) do you think could be omitted in a polarisation assay? How would a standard curve for such an assay appear?

The seperation (fifth step) can be omitted as this is now a non-separation assay. We merely mix the antibody, standards (or sample) and fluoro-tracer (Ag-F), incubate for the required time, and measure the polarisation of fluorescence in a polarisation fluorimeter. Fig. 6.2c shows the standard curve for such an assay. At zero concentration, the maximum number of tracer molecules are bound to antibody and hence the maximum signal is observed. As the concentration in the standard increases less tracer is bound and hence the polarisation signal falls.

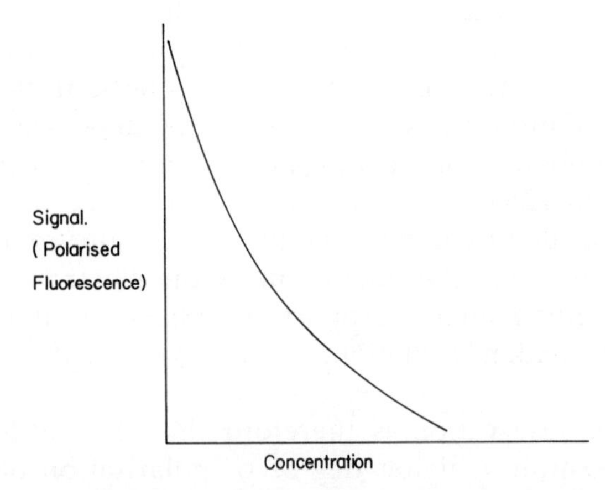


**Fig. 6.2c.** *Standard curve for a fluorescence polarisation assay*

**Current developments**

Several developments have been made to try to minimise the tendency of background fluorescence to produce an impoverished signal:noise ratio. The most obvious approach is to increase the number of fluorescing groups in the system. However there can be a limit to this since intramolecular interaction can lead to internal quenching. In systems in which the binding agent (usually an antibody) is labelled, this effect is minimised since the large size of the binder produces good spatial separation of the fluorescing groups. Incidentally another advantage of double antibody sandwich techniques is that it is possible to get up to twelve molecules of labelled second antibody attached to each molecule of first antibody giving the potential for a very strong signal.

An intensive search has been undertaken for fluorescent labels with a high quantum yield and a long fluorescent lifetime which would increase the strength and precision of measurement of the fluorescent signal. The use of phycoerythrin in the polarisation assay described above is an example of the results acrueing from the interests of commerce in these developments.

An alternative approach is to decrease the level of background fluorescence, and preferably by methods avoiding the need for sample purification procedures. An advantage of those systems in which a separation stage for free and bound label occurs is that the bound labelled ligand will be separated from the background fluorescing components (along with unbound labelled ligand) when it attaches to the fixed binding agent.

Another method for decreasing the background effect has been termed time-resolved fluorescence and takes advantage of the fact that under high light intensities, background fluorescence fades (bleaches) rather faster than the fluorescence of the label (Fig. 6.2d) giving a decrease in background of 3–4 orders of magnitude. While the actual fluorescent intensity at the time of measurement ($t_m$) is less than maximal, the signal: noise ratio is much improved. Short pulses of high intensity laser light have been used in these systems with a carefully chosen time delay, but unfortunately the timing of mixing and assay are rather critical for reliable results using the basic methods. The principle has however been developed into a more sophisticated system called (by LKB-Wallac (Pharmacia) who are the manufacturers), DELFIA – 'Dissociated Enhanced Lanthanide Fluoro-Immunoassay'. While these methods use the standard approaches described earlier, the labels are different in employing rare earth elements such as europium (Eu). For measurement, the europium can be released from bound states by special detergent solutions, and will form micelles with a fluorescent intensity of up to $10^6$ times that of uncomplexed atoms.

Both solid-phase competitive (separation) systems and solid phase sandwich (non-separation) procedures have been developed and in each case the signal is generated at the end of the equilibration reaction by the addition of the special detergent solution.

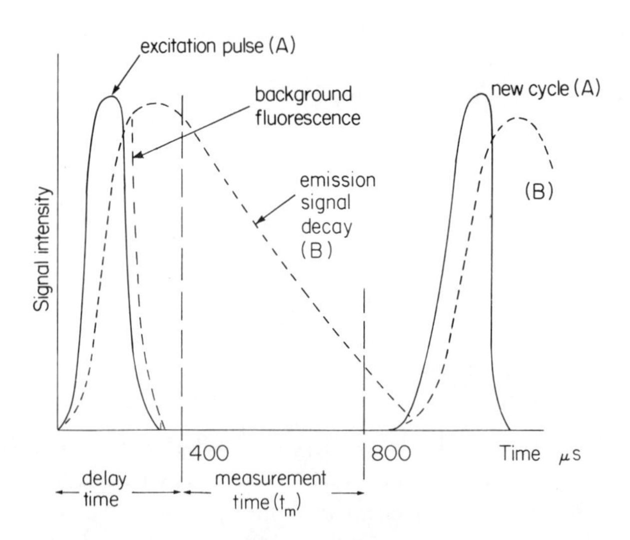


**Fig. 6.2d.** *Representation of the events involved in time-resolved fluorescence techniques*

The europium micelles also have other advantages, in particular:

- a large Stoke's shift; the difference between excitation (340 nm) and emission (613 nm) wavelengths, which minimises spectral interference;

- a broad excitation wavelength and a narrow emission wavelength, about 100 nm and 10 nm respectively. The former gives a high specific activity by producing a high excitation energy level, and the latter reduces background interference by allowing one to screen out non-important wavelengths;

- a long fluorescence decay time. Whereas the half-life for the decay of background fluorescence is of the order of 4 to 5 ns, that for europium is $10^3$ to $10^6$ ns, allowing the selection of a measurement window ($t_m$) which minimises the effect of the background fluorescence and gives a good margin for error in the timing;

- a high fluorescence intensity which generates low detection limits (theoretically of the order of $10^{-17}$ mol $dm^{-3}$).

The DELFIA assay for TSH (thyroid stimulating hormone) has, according to the manufacturer, the useful characteristics of a lower detection limit of $3 \times 10^{-5}$ international units $dm^{-3}$ (the adult range being 10–100 times this value), a linear response across two orders of magnitude of concentration and an acceptable precision of about 5%. It is also said to have such good specificity that in contrast to other immunoassays it actually gives a zero value for TSH depleted specimens !!

Methods are currently available for ferritin, alpha-fetoprotein (AFP), human chorionic gonadotrophin (HCG) and TSH. Assay times unfortunately can be up to 3 hours.

**SAQ 6.2b**

The abbreviated experimental protocols and data for two different DELFIA procedures are given.

(*i*) The assay of human luteinising hormone (LH).

- Pipette standards, experimental samples and europium-labelled anti-LH-antibody into tubes which are pre-coated with excess anti-LH.
- Incubate for 2 hours at room temperature.
- Wash off surplus reagents.
- Add the detergent solution.
- Measure the fluorescence.

| Concentration of standard. IU $dm^{-3}$ | Fluorescence $c\ s^{-1}$ |
|---|---|
| 1 | $2.8 \times 10^3$ |
| 5 | $1.5 \times 10^4$ |
| 50 | $1.7 \times 10^5$ |
| 250 | $9.1 \times 10^5$ |
| Sample | $9.0 \times 10^3$ |

(*ii*) The assay of cortisol.

- Pipette standards, experimental samples and europium-labelled anti-cortisol-antibody into tubes which are pre-coated with immobilised cortisol.
- Incubate for 2 hours at room temperature. ⟶

**SAQ 6.2b (cont.)**

- Wash off surplus reagents.
- Add to detergent solution.
- Measure the fluorescence.

| Concentration of standard nmol $dm^{-3}$ | Fluorescence c $s^{-1}$ |
|---|---|
| 100 | $7.0 \times 10^5$ |
| 250 | $4.8 \times 10^5$ |
| 500 | $3.5 \times 10^5$ |
| 1000 | $2.5 \times 10^5$ |
| 2000 | $1.8 \times 10^5$ |
| Sample | $4.0 \times 10^5$ |

- On the graph paper below, plot the data in each case and determine the concentration of LH or cortisol in the experimental samples. Plot sample concentration on the horizontal log axis.

- Explain the principles involved in these two procedures both of which are variants of the basic DELFIA system.

- When a sample containing 5 IU $dm^{-3}$ of LH and 5 IU $dm^{-3}$ of HCG (another of the gonadotrophin hormones) was assayed by method (*i*) the signal rose to $1.54 \times 10^4$ c $s^{-1}$. When a sample of 250 nmol $dm^{-3}$ cortisol and 250 nmol $dm^{-3}$ of corticosterone (which differs in lacking a 17-hydroxyl group) was assayed by method (*ii*) the signal obtained was $3.0 \times 10^5$ $cs^{-1}$. Explain this difference by considering the principles of the methods involved; you may find reference to the information in Section 5.5.1 useful.

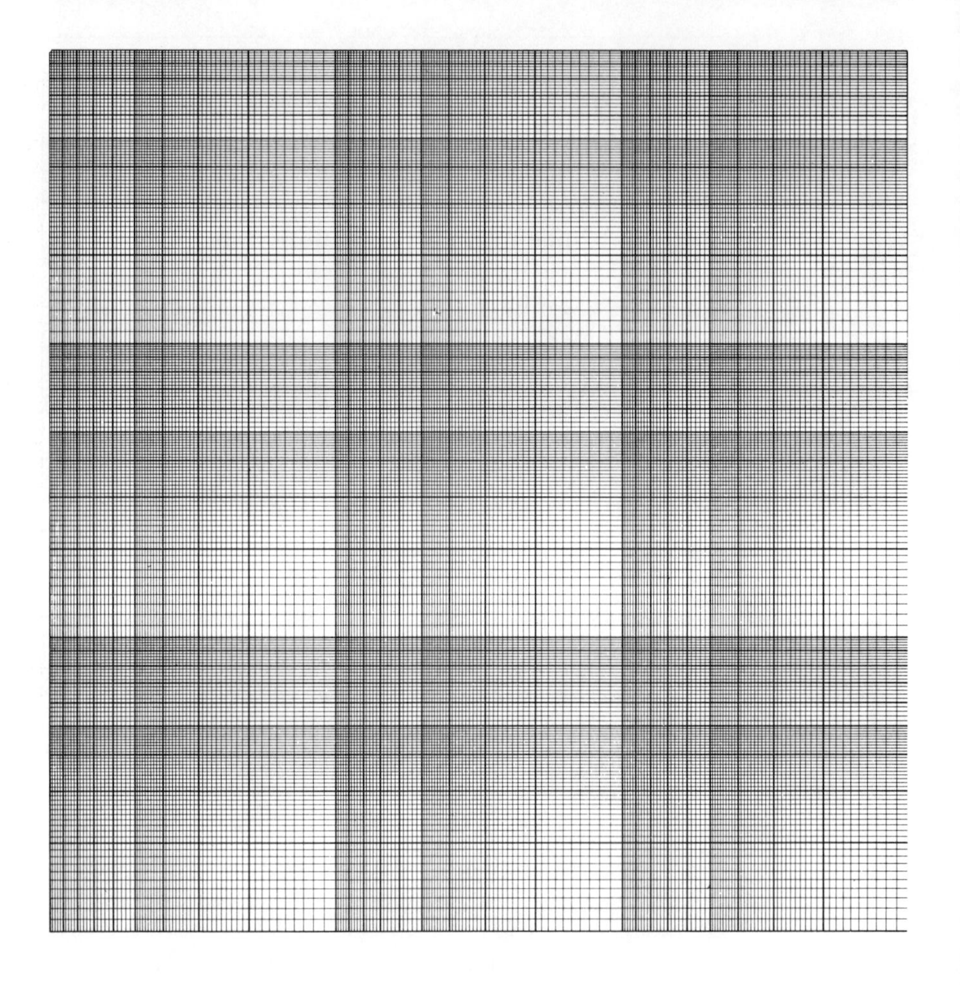

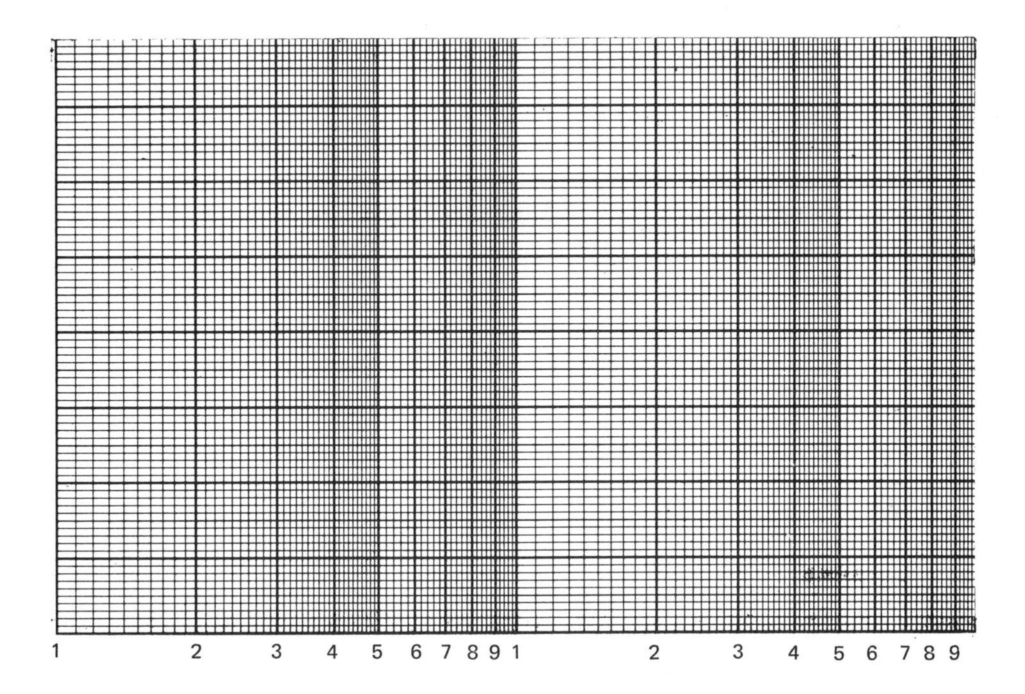


### 6.2.3. Luminescent labels

Luminescence is another process in which photons are emitted but in this case the energy for the excitation of the electrons comes from the cleavage of covalent chemical bonds. Luminol, for example, is readily oxidised by peroxidase or heavy metals to yield a 3-aminophthalate with the concurrent emission of photons (Fig. 6.2e).

$$\text{Luminol} + 2H_2O_2 + OH^- \xrightarrow{\text{peroxidase}} \text{3-aminophthalate }(NH_2,\ 2\,COO^-) + N_2 + 3H_2O + h\nu$$

Luminol

**Fig. 6.2e.** *Oxidation of luminol to yield radiation of wavelength 430 nm*

The reaction can easily be followed in any suitable photon counter but preferably in a specially designed luminometer. A wide range of substrates capable of taking part in chemiluminescent reactions are known and they can be used in the measurement of those enzymes (eg horse-radish peroxidase) used as labels in the so-called enzyme immunoassay systems that we shall deal with shortly. However the substrates can also be used as the labels themselves and luminol has been linked to antigens, haptens and to antibodies. In such simple systems the free and bound forms are separated and the appropriate phase placed in the luminometer. Under carefully timed conditions, sodium hydroxide, peroxidase and hydrogen peroxide are added and the light emitted during a defined period ($t_m$) is then measured (Fig. 6.2f). Non-separation systems are also available in which the binding reaction produces sufficient change in signal (ie of intensity or wavelength) for discrimination to be possible.

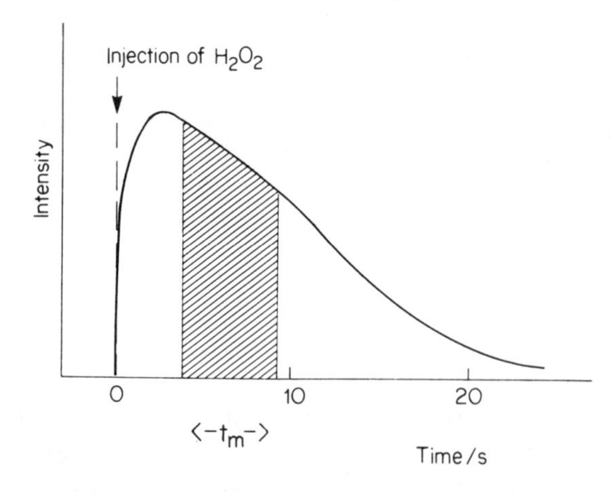


**Fig. 6.2f.** *Representation of the light signal obtained upon oxidation of a chemiluminescent moiety during treatment with alkaline hydrogen peroxide in the presence of peroxidase*

Unfortunately when used in this way luminol presents a number of problems, for example, the luminous intensity is in fact quite low when luminol molecules are coupled to other molecules. While derivatives such as the aminoalkylluminols, and bioluminescent rather than chemiluminescent labels are rather better in this regard, the recent introduction of acridinium derivatives is a significant step. These derivatives have a number of merits:

- they do not need a catalyst for chemiluminescence and partly as a consequence of this, are less subject to interference and spontaneous reactions due to components of complex biological samples,

- the oxidation occurs rapidly ($<1$ s) which leads to the possibility of short assay times,

- the luminescent species only appears following its release during the reaction, coupled labels give no chemiluminescence; consequently a high signal: noise ratio is obtained,

- acridinium-labelled materials tend to have a high specific activity, often giving over $10^{15}$ photons per gram of antibody, compared with say $^{125}$I labelled antibody which may have an average of only about $5 \times 10^{11}$ $ds^{-1}$ per gram of antibody.

However, there are disadvantages associated with using acridinium esters, such as the inherent instability of the phenyl ester bond linking this antigen to the acridinium nucleus (Fig. 6.2g(*ii*)).

Several systems are available, some of which are marketed by the firm of Ciba Corning Diagnostics under the trade name of MAGIC LITE. In the most straightforward of these systems (Fig. 6.2g (*i*)), sample antigen ($Ag_S$) competes with immobilised antigen for acridinium labelled antibody Ab-AC, and after equilibration and separation some of the label will be in the form

$$\text{support} - \text{Ag} - \text{Ab} - \text{AC}$$

and the concentration of bound labelled antibody will be inversely proportional to the original sample antigen concentration. Measurement of the label follows the addition of sodium hydroxide and peroxide, since under alkaline conditions the peroxide attacks the acridinium ester link to the antibody releasing an excited tricylic compound such as 10-methylacridone Fig. 6.2g (*ii*). A high reaction rate contributes to the generation of an intense flash of light and a method with very low detection limits.

(i)

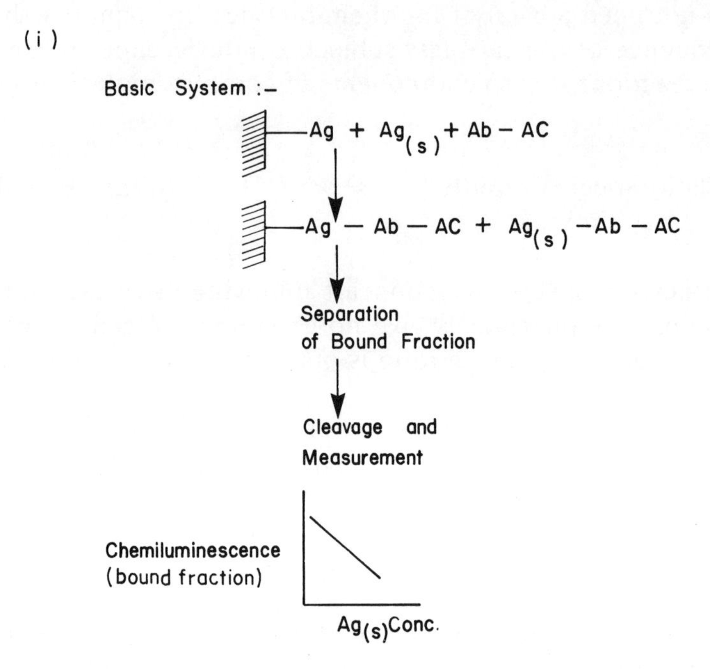


(ii)

$CH_3$ N$^+$ CO O R

$H_2O_2$ / $OH^-$

$CH_3$ N O

An aryl acridinium ester

10 — Methylacridone

**Fig. 6.2g.** *(i) Representation of a simple acridinium (AC) labelled competitive chemiluminescent immunoassay system. (ii) The reaction of arylacridinium esters*

The first series of assay kits produced by the manufacturer was for the investigation of thyroid gland related disorders by measurement of thyroid stimulating hormone (TSH), total and free thyroxine ($T_4$), and triiodothyronine. Other kits will undoubtedly be available in the near future.

Π For the following assays does Fig. 6.2h represent a plot of antibody bound or unbound (free) label against concentration of the standard?

(*i*) Heterogeneous competitive radiolabelled ligand assays.

(*ii*) Enhancement fluorescence assays.

(*iii*) Indirect quenching fluorescence assays.

(*iv*) Fluorescence polarisation assays.

(*v*) Chemiluminescence assays using acridinium labelled antibody.

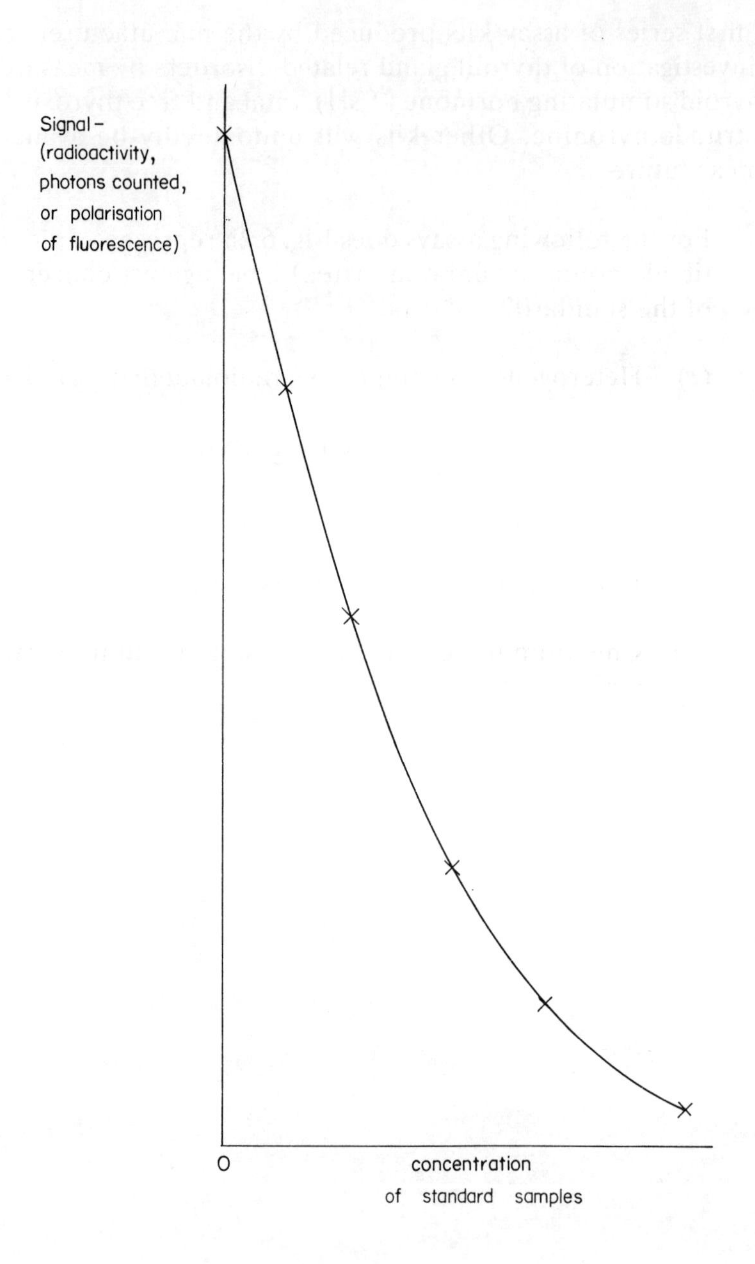


**Fig. 6.2h.** *Plot of the signal produced in various assay systems against the concentration of standard samples*

The answers, which you should have given are as follows:

(*i*) bound label;
(*ii*) bound label;
(*iii*) free label;
(*iv*) bound label;
(*v*) bound label.

It is perhaps the case that systems using photon-emission labels are more diverse and complex than any other aspect of immunoassay, and more detail can be obtained from the references cited in the Bibliography.

### 6.2.4. Enzyme labels

In the basic luminescence assay we added an enzyme (peroxidase) to the antigen/luminol conjugate to produce photons as a measurable signal. If instead we had an antigen/enzyme conjugate in the assay we could use the enzyme activity as the signal. The distribution of the enzyme in bound and unbound fractions will, as before, be related to the concentration of competing agent in the original incubation mixture. Hence after separation of the free and bound forms, substrate can be added, and the rate of reaction determined in one of the fractions, that rate being dependent on the amount of enzyme in that fraction.

Ideally an enzyme is chosen with a high turnover rate and which produces a product which has easily measured properties, such as a high extinction coefficient or measurable fluorescence, from a substrate giving a minimal signal (ie no absorbance or fluorescence). When considerations of cost, enzyme stability and resistance to interference by sample components, and the need to ensure minimal interference with the fundamental immunological reactions of the labelled molecules are taken into account, most methods use either horse radish peroxidase, alkaline phosphatase, galactosidase or glucose-6-phosphate dehydrogenase. This general procedure is termed an Enzyme Linked Immunosorbent Assay (ELISA), and the steps involved in one format are described below and shown in Fig. 6.2i(*i*).

The enzyme-labelled antigen, Ag-E, and a sample or standard containing unlabelled antigen, Ag(s), are added to tubes or wells of plastic microtitre plates on the surface of which is adsorbed the specific antibody for that antigen.

The antigen molecules compete for, and bind to, the antibody in the same proportions as they were present originally, and unbound components can be removed from the tubes by decantation and washing.

Substrate is now added to the tubes and the product appears at a rate dependent on the enzyme activity in the bound fraction. The signal (reaction rate) is therefore inversely related to the concentration of unlabelled antigen present in the sample or standard.

It has also proved possible to develop non-competitive methods using these principles Fig. 6.2i(*ii*). One such involves the binding of large (ie excess) quantities of antibody to a fixed surface. On the addition of sample all the corresponding antigen molecules will bind and be retained. The quantity of bound antigen is determined by the addition of a second, reagent antibody carrying an enzyme label. The reaction rate will increase as the concentration of antigen rises. This technique then is a two-site sandwich immunoenzymetric assay with the advantage of high specificity through the use of two antibodies, but with the disadvantage of high cost.

The majority of ELISA assays employ immobilised antibodies since it is usually necessary to introduce a washing step to remove materials present in the complex biological samples that are being analysed, in order to prevent them interfering with the enzyme assay that generates the signal. Immobilisation of antibody on a surface enables easy separation and washing to be undertaken.

The value of ELISA assays can usefully be illustrated by reference to the measurement of the hormone HCG which is produced in large quantities by the placenta and by certain tumours. The necessity for early diagnosis of the latter by measurement of low levels of HCG is perhaps self-evident, but sensitive tests able to follow the remission of a cancer to its completion are also important. Confirmation and monitoring of early pregnancy in cases of poor fertility or troublesome pregnancy (eg a tendency to spontaneous abortion) is a more frequent requirement in health care than is often realised.

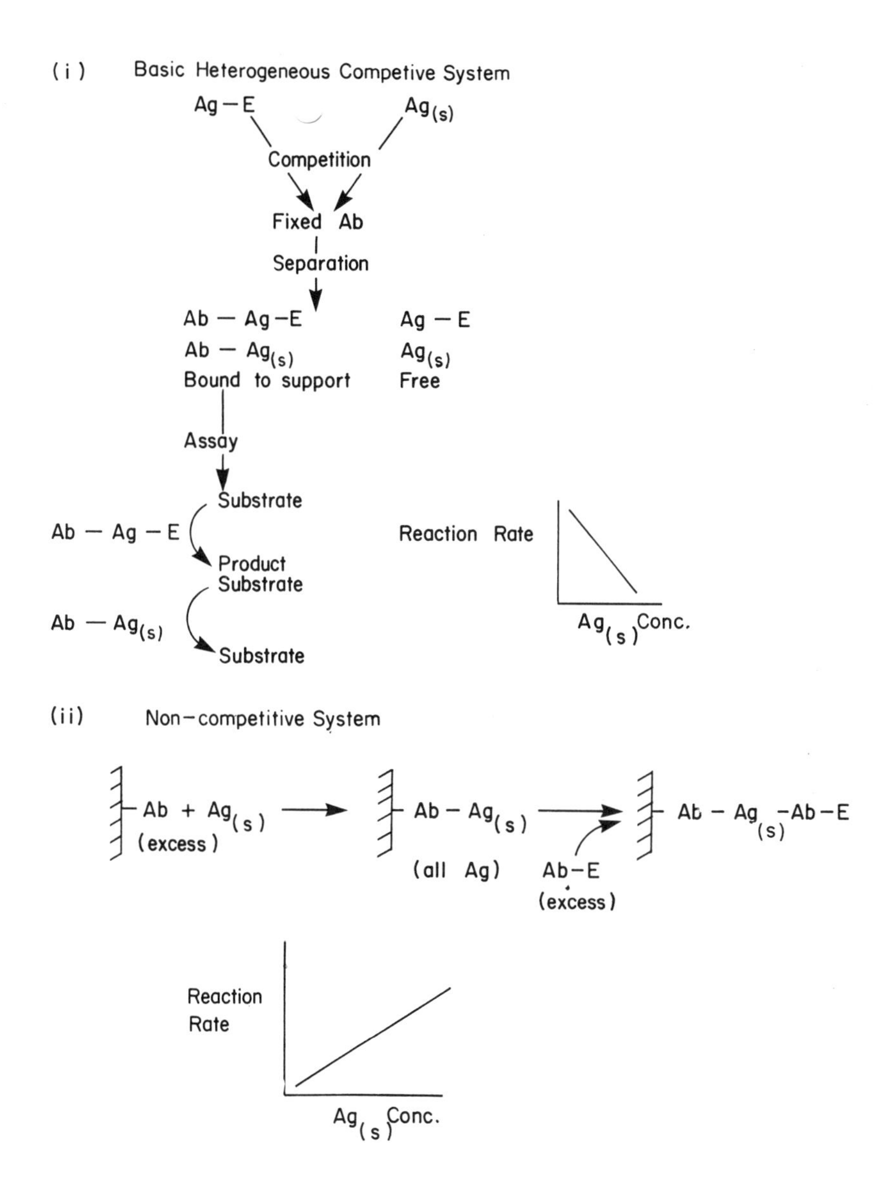


**Fig. 6.2i.** *Representation of typical enzyme linked immunosorbent assays (ELISA)*
*(i) Basic heterogeneous competitive system*
*(ii) Non-competitive system*

Commonly-used agglutination immunoassays for HCG can only detect levels of HCG down to about 200 IU $dm^{-3}$ and in practice therefore only reliably detect pregnancy some days after the first missed period. Although ELISA assays with a minimum detectable concentration of about 5 IU $dm^{-3}$ are not as good as IRMA assays, with a limit of 3 IU $dm^{-3}$ they have nonetheless proved able to detect pregnancies 7–10 days after conception and long before the first missed period.

However since 5 IU HCG $dm^{-3}$ is thought to correspond to perhaps $5 \times 10^6$ tumour cells, there is still a need for further improvement in these assays.

A distinct advantage of ELISA techniques is the production of a visible signal. Many qualitative or semi-quantitative methods for the simple demonstration of the presence of materials have been developed using visual assessment of the end product, particularly in the fields of microbiology and eukaryotic parasitology. ELISA techniques have the advantages of using a cheap, safe and non-licence requiring label, being fairly simple, amenable to semi-automation and consequently being particularly suitable for large scale screening and in financially restricted countries and organisations. Interestingly a significant number of assays in these fields are for antibodies produced in response to previous or current infections rather than for the cell surface antigens of the organisms themselves. A wide range of procedures for temperate or tropical bacterial diseases (eg brucellosis, cholera, yaws), viral diseases (eg rubella, influenza, gastroenteritis) or parasites (eg toxoplasmosis, toxocariasis, malaria) are available. Disease surveillance in domesticated animals and the screening of blood and blood products for hepatitis B and AIDS virus are other important applications.

In the ELISA technique described above, enzyme/antigen conjugates and enzyme/antibody conjugates were employed in what should by now be fairly familiar approaches to immunoassays. However if a large molecular mass molecule was present on the surface of the enzyme adjacent to the enzyme's active site, this could block the approach of the substrate to the active site, and we could use this restriction to develop an assay based upon inhibition of enzyme activity by the binding reaction. A small hapten (H) moiety or antigen is attached to the labelling enzyme near its active site. While this is too small to block the approach of the substrate to the active site under normal circumstances, when a large antibody molecule binds to it, access to the active site is prevented and the enzyme reaction inhibited. (Fig. 6.2j (*i*)). After the binding reaction only the unbound hapten enzyme conjugate will be able to catalyse the reaction Fig. 6.2j(*ii*). These assays are therefore a little unusual in measuring the free fraction rather than the bound one.

This phenomenon forms the basis of a homogeneous enzyme immunoassay method called the Enzyme Multiplied Immunoassay Technique (EMIT – a registered trademark of the Syva Corporation). The technique has the merit of being a non-separation method, in which one merely mixes the antibody with a mixture of enzyme/hapten conjugate and sample hapten, and allows binding to occur. As the concentration of sample hapten rises, less of the limited number of antibody molecules will be available to bind the enzyme conjugate leaving more of the latter to catalyse the reaction. The signal is generated by unbound labelled hapten and is therefore directly related to the original analyte concentration.

Another major advantage of the EMIT principle is speed. Syva assays for the drugs phenobarbitone and diphenylhydantoin take less than 2 minutes per tube, allowing a series of standards and experimental samples to be measured in less than 15 minutes from receipt of the (possibly emergency) specimen. Many current RIA methods still take an hour or more for completion. It is unfortunate that procedures have not been developed for more analytes.

**SAQ 6.2c**

In one type of ELISA assay a limited amount of antibody is bound to a surface and enzyme-labelled antigen, and sample containing antigen is added. In an alternative type an excess of antibody is bound to the surface, the sample is added and this followed by an excess of enzyme-labelled antibody to generate a signal.

Which of the above two systems is non-competitive?

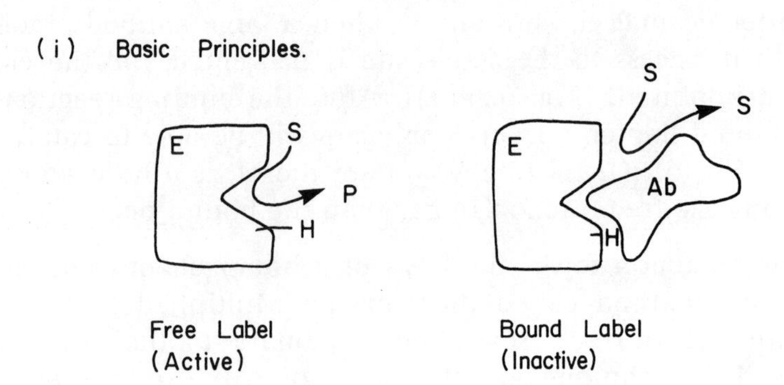


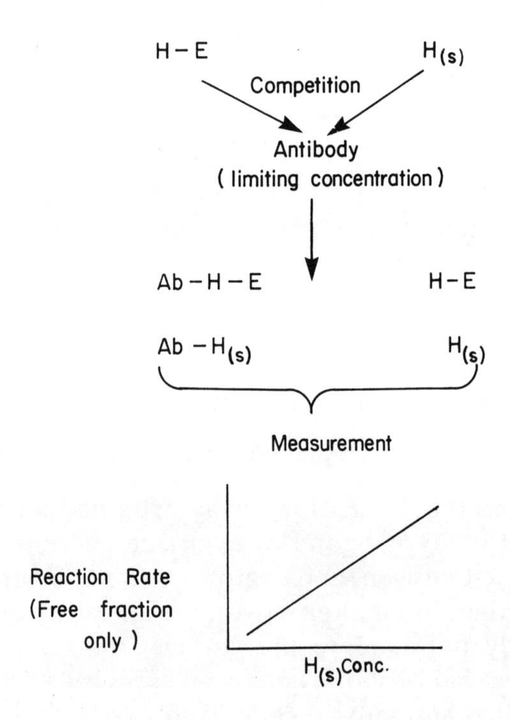


**Fig. 6.2j.** *The enzyme multiplied immunosorbent assay technique (EMIT)*

**SAQ 6.2d**

In a typical immunoenzymetric assay three steps are involved in which a sequence of attachments occur. Describe or illustrate diagrammatically these three attachment steps. Is this a competitive or non-competitive assay?

The three label systems so far described are perhaps the most commonly used, and a summary of their relative merits might be useful. This is given in table form, Fig. 6.2k, with a lower number indicating a more satisfactory characteristic.

It should be noted that this table represents opinions which are at least partly, and sometimes completely, subjective.

| Assay Feature | Assay Label | | | Problems |
|---|---|---|---|---|
| | Radio-isotope | Fluorescence | Enzyme | |
| Label stability | 3 | 1 | 2 | Enzymes denature, radio-isotopes decay |
| Stability of labelled material | 2 | 1 | 1 | Radiolysis is possible |
| Effect of label on properties of antigen or hapten | 1 | 2 | 3 | Enzyme molecules are huge, fluorescent ones generally much smaller. |
| Specific activity | 1 | 1 | 2 | Multiple labelling is possible for fluorescence and radio-isotope labels, but internal quenching may occur with fluorescence. Strong signals from enzyme labels arise from catalytic cycling. |

| | | | | |
|---|---|---|---|---|
| Detection limits | 1 | 2 | 3 | |
| Sensitivity (slope) | 1 | 1 | 2 | |
| Accuracy and precision | 1 | 1 | 2 | |
| Safety | 2 | 1 | 1 | Enzyme substrates may be toxic, radiation is dangerous but levels are low |
| Assay time | 3 | 2 | 1 | The EMIT procedures can be as short as 2-4 min. |
| *Label measurement* | | | | |
| complexity of theory | specialised | may be specialised | generally easy & known | |
| principles of measurement techniques | only one but can be complex | a very wide range of techniques | | |
| instrumentation | specialised | may be specialised | generally available | |
| signal : noise ratio | high | lower due to background | high | |

**Fig. 6.2k.** *A comparison of the relative merits of radioisotopes, fluorescent and enzymic labels in immunoassay (a lower number indicates a more satisfactory characteristic)*

### 6.2.5. Assays using substrates as labels

In the system shown in Fig. 6.2j it should be possible to link the substrate (S) rather than the enzyme (E) to the hapten or antigen. An antibody to the H-S conjugate could be added and the rate of product formation following the addition of unconjugated enzyme would be measured as the signal.

Π Which of the following do you think would happen after the addition of antibody to the H-S conjugate?

(*a*) the rate of S→P reaction would decrease,
(*b*) the rate of S→P reaction would increase,
(*c*) there would be no change in the rate,
(*d*) the reaction would not occur.

The answer is (*d*). On binding of the H-S conjugate to the antibody the substrate would be unable to fit into the enzyme's active site; hence no reaction (substrate to product conversion) could occur. This principle forms the basis of another commercial, competitive heterogeneous immunoassay method known (since the product in this case is fluorescent) as Substrate Labelled FluorImmunoAssay (SLFIA).

One of the main advantages of using enzymes as labels in immunoassays is that the catalytic nature of enzyme activity ensures that the signal they produce increases with time, assuming that the substrate does not become depleted or the enzyme adversely affected in any way. With EMIT assays for example, competition occurs between enzyme labelled hapten and unlabelled hapten for a limited supply of antibody, and excess substrate may be added to generate an ever increasing signal from the remaining unbound enzyme-hapten complex (Fig. 6.2l (i)). In the case of SLFIA, however such amplification does not occur (Fig. 6.2l (ii)) since a fixed, limiting amount of hapten-substrate conjugate is employed to compete with the unsubstituted hapten for antibody in a typical homogeneous assay and each conjugate molecule will only give one product molecule for the signal.

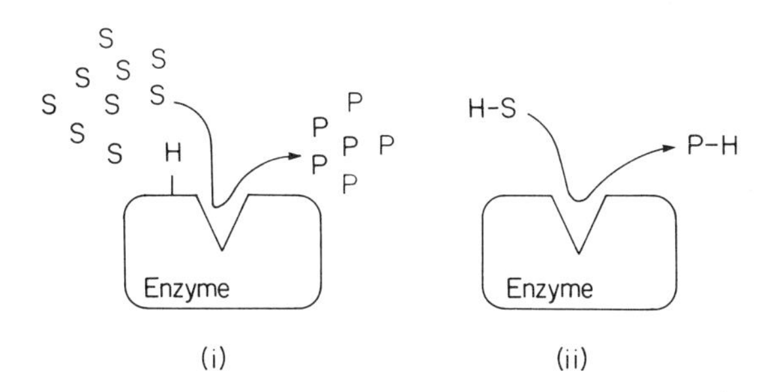


**Fig. 6.2l.** *A comparison of the efficiency of enzyme-labelled (i) and substrate-labelled (ii) hapten in the generation of a measurable signal (reaction product)*

**SAQ 6.2e**

> Which assay method (EMIT or SLFIA) do you consider likely to have the lower detection limits? Explain the reasons for your choice.

### 6.2.6. Less common labels

The tremendous importance of immunoassay techniques has led workers (many employed by commercial companies) to search for alternative approaches and variants on the basic techniques, in the hope of finding methods of particular merit for certain applications, or indeed of general value, or for commercial and financial reasons.

Two such alternative less common labelling systems are described below to illustrate this point.

(*a*) The AMES Corporation market an ARIS system (Apoenzyme Reaction Immunoassay), which uses an enzyme component (prosthetic group) referred to as FAD as the label, and an FAD-deficient glucose oxidase enzyme as a measuring system. Thus Ag bound FAD is capable of reactivating the enzyme which in turn will produce a measurable signal in the assay. Antibody binding prevents the FAD/glucose oxidase interaction and the extent of antibody binding is therefore reflected in a decrease of the glucose oxidase generated signal. Kits are already available for a number of antibiotics and anticonvulsants, and for the important antiasthmatic drug theophylline. The fact that AMES have managed to build this principle into a dry-film (strip) assay in which a coloured reaction product is measured by light reflectance at 740 nm could make it another technique of considerable usefulness in the future.

(*b*) Relatively large particles have been used for labels in a number of systems. One such involves 50 nm gold particles carrying appropriate antibodies, and since a dispersal of such particles gives a sol, the technique has been given the name of Sol Particle Immunoassay (SPIA) (Diagnostics Research Laboratories, Belgium). The technique is said to have the advantages of being non-isotopic, with a high reagent stability, a high precision due to the use of non-chemical detection methods, and low detection limits of $10^{-13}$ to $10^{-14}$ mol dm$^{-3}$ using atomic absorption spectroscopy as an assay for the particles. The magenta colour of the colloidal end product can also be measured by visual or spectroscopic means with detection limits of $10^{-11}$ to $10^{-12}$ mol dm$^{-3}$.

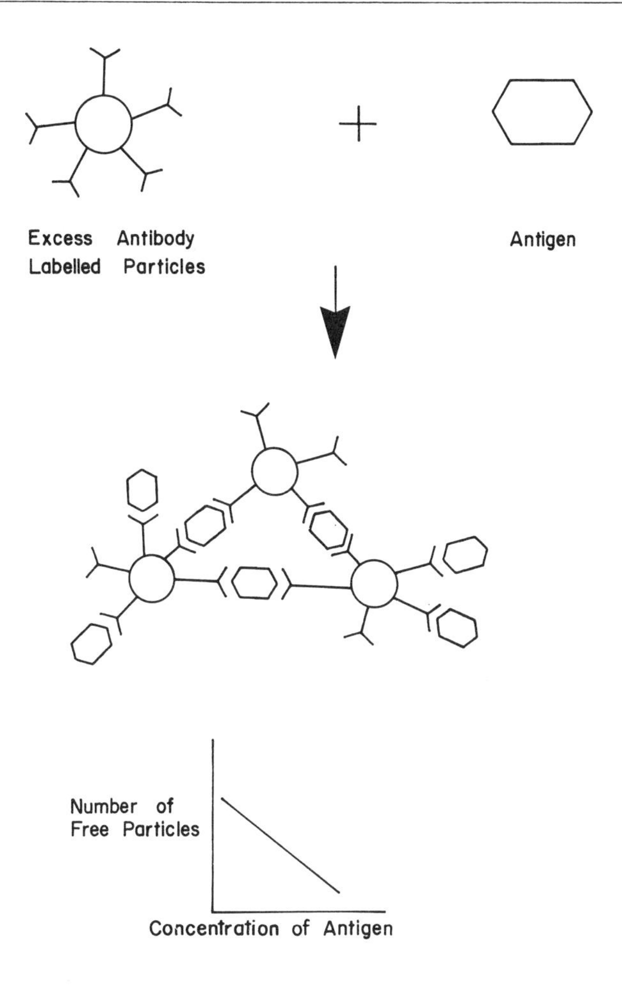


**Fig. 6.2m.** *Diagrammatic representation of the use of antibody labelled particles in an immunoassay system*

Another novel feature of this type of system is that perhaps because the antibody molecules are clearly exposed, the antigen will bind to the antibody relatively easily and will cause sufficient cross-linking between the antibody labelled particles to give substantial precipitation (Fig. 6.2m).

Measurement can therefore be carried out by means of an electronic particle counter with a window set to measure the unagglutinated remaining particles. The agglutinated particles will be of variable size and less easy to measure reliably. A similar and cheaper system based on the use of 0.8 $\mu$m (polystyrene) latex particles has been marketed under the trade name of IMPACT

Particle techniques have some definite merits including being fast and automatable, with reagents said to be stable for at least 2 years and with very low detection limits (in general between $10^{-15}$ and $10^{-18}$ mol dm$^{-3}$). The most significant problem with the methods is the non-specific particle agglutination caused by proteinaceous materials such as the Clq component of complement and the Rheumatoid Factor present in blood. It seems possible that partial degradation of antibody molecules (to exclude the so-called Fc fragment) might remove the binding site for these proteins while retaining the specific site for the corresponding antigen or hapten.

**SAQ 6.2f**

Illustrate the end product of a particle-based sandwich heterogeneous (separation) technique for the assay of an antigen using a fixed primary antibody.

Use the following symbols for the different components of the system:

support ⦚
antibody Ab
antigen Ag
particle labelled with antibody Ab-P

## 6.3. SEPARATION PROCEDURES

Many immunoassay procedures (particularly of the older types) require the separation of free and bound forms of label before measurement, and a range of techniques has been evolved to do this. An ideal separation procedure should be simple, fast, highly reproducible, applicable to a wide range of antigens and allow the use of the same tube for incubation, separation and measurement. Furthermore it should not be affected by the volume or protein concentration of the mixture or disturb the equilibrium or kinetics of the reactions in the assay system. Some of the procedures are now out-of-date and are given at the beginning of the list below for historical interest and completeness. The others have found applications in different systems, and since several have been mentioned in this text before, this list is therefore largely a summary for reference purposes.

- Proteins are involved in immunoassays as antibodies if not as antigens, so standard protein precipitation techniques can and have been used, especially those employing salts and solvents.

- In some cases the antigen/antibody complex is sufficiently large or otherwise unstable to precipitate spontaneously, enabling a separation by centrifugation or possibly filtration. This precipitation can be made more complete, reproducible and rapid by using a precipitating second antibody (anti-immunoglobulin) raised in a different animal from the first antibody and with the latter acting as antigen. The end result could be for example

  $$\text{Ag-Ab}_1\text{(eg rabbit)-Ab}_2\text{(eg goat)}.$$

  Polyethyleneglycol (PEG) is often added to obtain a precipitation with a low number of molecules in the complex to give a marked increase in speed and specificity to the process.

- Some proteins, especially antibodies, bind spontaneously and quite strongly to glass or plastic supports. They can therefore be attached to beads, tubes, columns or wells of microtitre plates allowing non-binding materials to be removed as a centrifugation supernatant, by decantation, or by elution followed by rinsing. The need for centrifugation introduces delays in the assays whereas the coating of tubes and particles reduces the reaction rate due to immobilisation of the antibody. Also coating may not be reproducible, may be limited in extent and antibodies may detach during the latter stages of an assay.

- A novel system marketed by Serono Diagnostics makes use of cellulose particles with an iron(III)oxide core. The antibodies can be attached to the cellulose and the particles retained within the tubes during a rinsing and decantation procedure, by making use of the magnetic properties provided by the iron(III)oxide. Several commercial systems are now available making use of a number of different magnetic components and particle chemistries (agarose, polyacrylamide gel, latex and even albumin). Assay kits for thyroxin ($T_4$), human placental lactogen (HPL), and digoxin are among those marketed. The systems have the advantages of small particle size giving paramagnetic properties, a reluctance to sediment spontaneously, a large surface area and covalent (hence stable) linking of antibody.

**SAQ 6.3a**

Briefly describe how the separation of bound labelled antigen can be achieved in the three systems (A, B and C) shown in Fig. 6.3a.

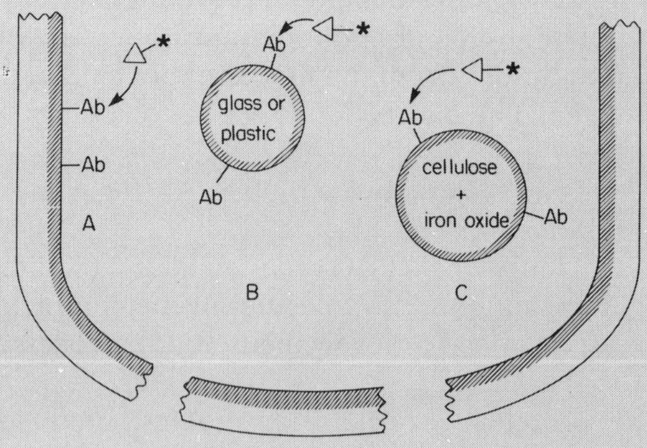

**Fig. 6.3a.** *Separation systems using antibody bound to various supports*

**SAQ 6.3a**

It ought to be noted that the tendency has been to develop non-separation assays wherever possible. As a general rule the more steps that are required in any assay the greater the risks of mistakes leading to inaccuracies and imprecision, the longer the assay time and the greater the cost. Separation techniques leading to incomplete separation introduce significant error due to some of the free fraction behaving like or being measured with the bound fraction or *vice versa*. Another very significant point is that it is more difficult to automate techniques requiring a separation procedure, which makes non-separation processes even more desirable. The serious impediment of a common requirement for high antibody concentration in non-separation procedures has been reduced with the advent of monoclonal antibodies available at reasonable cost.

## 6.4. CHOICE OF METHOD

Π You might care to consider the factors upon which the choice of a method depends before reading the list below (which is of course by no means exhaustive).

The choice of method to be used will depend on a variety of factors.

(*a*) The number of samples to be processed in a batch and the number of assays to be performed each year.

(*b*) Whether an automated assay will be necessary or desirable.

(*c*) The speed of each assay and the necessity or desirability, on either clinical grounds or because of sample stability problems, to have a rapid assay.

(*d*) The amount of sample available for each assay and the effects of the matrix in which the analyte is to be measured.

(*e*) A need to analyse more than one component in each sample.

(*f*) The cost of the assay reagents and the cost of the labour (a product of the time involved and the level of skill required) to perform the assays.

(*g*) The availability and the cost of the instruments used to carry out the assay (particularly the measurement of the final signal).

(*h*) The necessary specificity, accuracy and precision.

(*i*) The detection limits of the method in relation to the likely concentration of analyte to be measured.

(*j*) The required sensitivity of the method (ie its ability to differentiate samples with similar concentrations), bearing in mind the expected range of concentrations in pathological and normal specimens.

(*k*) Whether a qualitative or a quantitative analysis is required.

## 6.5. FUNDAMENTAL STUDIES AND THE EVALUATION OF AN IMMUNOASSAY PROCEDURE

While developing an immunoassay procedure it is necessary to undertake some basic investigations of the procedure, especially of the immunological aspects of it, and also to evaluate its analytical characteristics (accuracy, precision etc) in order to determine its usefulness and limitations. This is in addition to studies of fundamental practical aspects of the procedure such as the determination of optimum incubation times for binding to reach equilibrium, optimum incubation temperature, and the choice of an appropriate buffer system. For a separation method the selection of a suitable separation system and its conditions will also need to be considered.

Critical to the development of a method is a knowledge of the required sensitivity, detection limits, specificity, accuracy and precision for a particular application. It is important to realise that in a routine analytical laboratory no effort is made to produce a 'perfect' method. One that is adequate to provide the necessary information is quite acceptable. This important point of philosophy commonly means that relatively simple techniques and methods are employed in clinical laboratories when more complex (but more expensive or skill-demanding) methods are available, perhaps even within that laboratory.

Bearing these points in mind it is perhaps not surprising that commercially manufactured kits are so commonly used in these investigations. A number of laboratories, however, have reported that it is unwise to rely entirely on the manufacturer to control the quality of the materials and to eliminate the problems in the methods he markets. Hence in common with other assays, on-going monitoring will be necessary to ensure that these parameters do not deteriorate as the assay is used over a period of time and staff become familiar with the method and perhaps less careful, reagents and apparatus age, and possibly the samples change in nature.

### 6.5.1. Some potential immunological and physiological problems

There is a very large number of possible problems deriving from the complexity of the physiological and immunological processes involved in these methods, and anyone developing a new method needs to answer the following questions.

(*a*) Does the labelled antigen behave qualitatively and quantitatively in the assay system like unlabelled antigen? This can be especially important if a large label (eg an enzyme) is used, or if the coupling procedure requires rigorous chemical reactions. This possibility is commonly investigated by producing a standard curve (Fig. 6.5a(*i*)) with a constant concentration of labelled antigen, a constant low concentration of binder (antibody) selected to give 50% binding of labelled antigen, and a variable concentration of unlabelled antigen.

If the experiment is repeated without unlabelled antigen, and with varying concentrations of labelled antigen, an identical slope for the measurement of free label should be obtained if the labelled and unlabelled forms react the same quantitatively. (Fig. 6.5a(*ii*)).

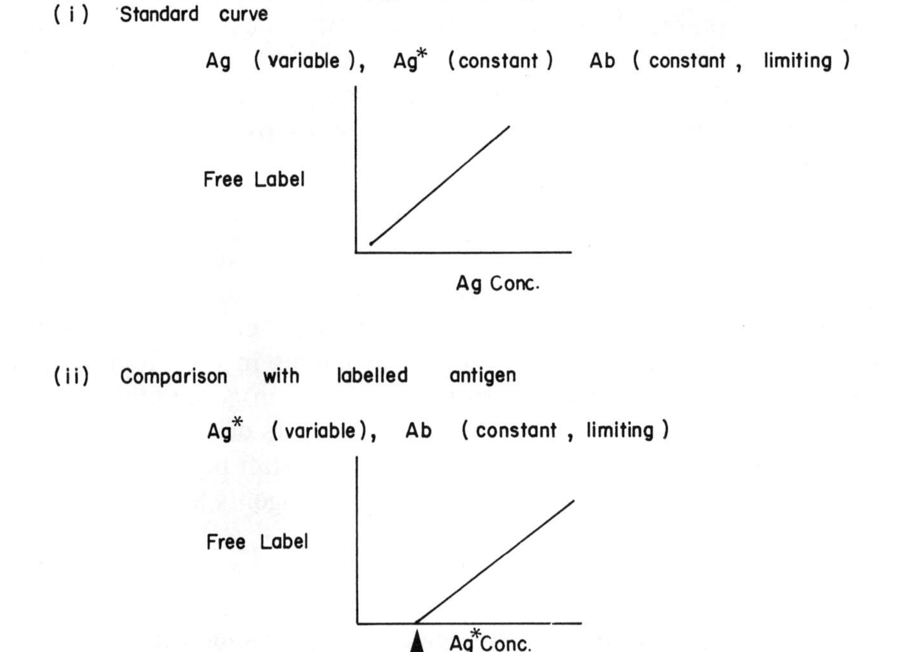


**Fig. 6.5a.** *A comparison of the reactivity of labelled and unlabelled antigen*

(*b*) Does antigen in the sample behave qualitatively and quantitatively like a pure standard sample of antigen? Many components of biological samples, independently or collectively, are capable of influencing chemical and binding activities (a so-called matrix effect), and many analytes (hormones, drugs etc) bind to carrier proteins and may not therefore be so readily available for other reactions. Occasionally artefacts can be introduced by degradation since peptides and proteins are subject to proteolytic damage caused by enzymes in biological samples. Differential damage to labelled antigens in standards, antigens in experimental samples and reagent antibodies can occur. It is unfortunately not always valid to assume that when an antibody and experimental sample are incubated the interfering enzyme might degrade both in parallel since the simpler chemistry of smaller peptides and haptens may make them resistant to proteolytic attack. The hormone gastrin for example is resistant to trypsin attack due to the absence of the amino acids arginine or lysine in its sequence, these being required for binding and cleavage by trypsin. It is unlikely that molecules of anti-gastrin would be similarly resistant.

Since most of the assays will be performed on biological samples, and such samples vary in their composition, it is necessary to demonstrate the extent of these effects. They are particularly important in systems which do not use radioactive labels; the colour or fluorescence properties of urine and serum can easily interfere with spectroscopic determinations of end-points. Additionally, materials may be present in the sample which partially or completely inhibit, or even on occasions, stimulate enzyme activity, and of course almost any system can be subject to the proteolysis problem described above.

The presence of some matrix effects can be investigated by the recovery (or known addition, or internal standardisation) technique where a known quantity of pure standard is added to a biological sample and the actual change observed compared with that expected from a knowledge of the amount added. An alternative approach is to measure the activity in a so-called zero standard, from which the target analyte is absent. With the advent of the highly specific technique of affinity chromatography, the selective removal of certain compounds from biological samples is now more practicable. The presence of significant proteolysis can sometimes be shown by including in the assay a control tube of labelled antigen and sample with the antibody being added after rather than before the normal incubation period for the assay. A reduction in the ability of the labelled antigen to bind the antibody is significant.

(*c*) What is the specificity of the antibody for the antigen and indeed for other aspects of the system in which binding occurs? Do molecular precursors, degradation fragments, molecular aggregates or similar compounds react and to what extent?

An important point is that the coupling of a hapten to a carrier molecule may have produced an immunogen that generates an antibody which is specific to that complex but reacts poorly or unspecifically with the unconjugated sample.

These points have been alluded to earlier in our considerations of the basic immunology involved in these assays but they are of great significance in biological applications of the assays because many physiologically active molecules also have precursors (insulin has pro- and even pre-pro- forms), and degradation fragments (eg the so-called C-peptide deriving from the conversion of pro-insulin to insulin). In Section 5.5.1 we considered the fact that a number of very important hormones (collectively called gonadotrophins) are all dimers of one $\alpha$- and one $\beta$-unit, and for TSH, LH, FSH and HCG the $\alpha$-units are unfortunately identical making it necessary to use antibodies which recognise determinants on the $\beta$-unit. One significant advantage of two-site immunoassays for these hormones is that the use of two antibodies greatly increases the specificity for these hormones.

Likewise steroid hormones have a common tetracyclic nucleus, but differ in the substituents they carry on this nucleus. However if the different substituents of these haptenic molecules are at positions adjacent to the bridge linking the hapten to the carrier protein, cross-reactivity may again result, due to a failure of the immunological system to recognise the differences between the immunogens when antibody is being raised.

Consider dexamethazone (I), prednisolone (II) and cortisone (III), the structures of which are shown in Fig. 6.5b. If the immunogen bridge is located at position 3 in the nucleus, the various substituents on the D ring will be exposed, leading to the production of different antibodies and an assay which is capable of distinguishing between I, and the other two (since I has an additional methyl group). If the

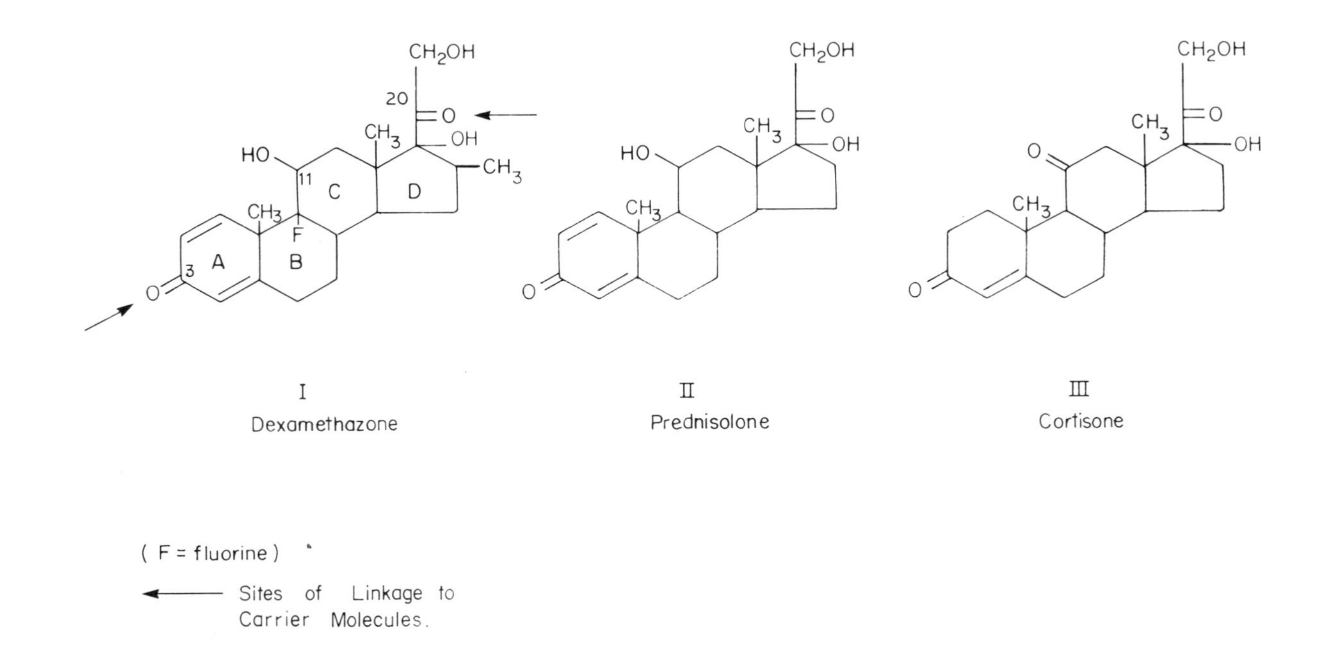


**Fig. 6.5b.** *The distinction between three steroid compounds using immunogenic carriers linked to different sites*

bridge is through the C-20 position, the assay should distinguish III from the others since it has a different ring A structure from I and II.

**SAQ 6.5a**

Barbiturate drugs are commonly coupled to an antigenic carrier via their group on the C-5 position. If pentobarbitone (Fig. 6.5c) is used to generate an antibody, which of the compounds A (barbitone), B (phenobarbitone) or C (quinalbarbitone), do you think might cross-react with the resultant antibody?

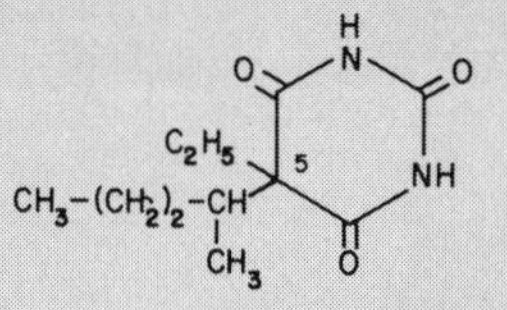

Pentobarbitone

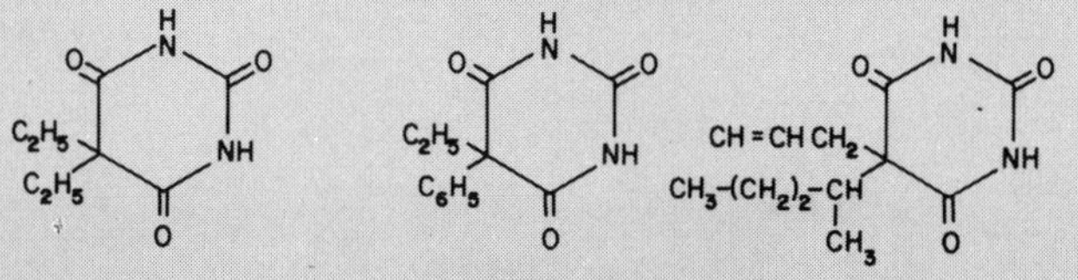

A Barbitone

B Phenobarbitone

C Quinalbarbitone

**Fig. 6.5c.** *A comparison of the structures of some barbiturates*

**SAQ 6.5a**

(*d*) An interesting physiological problem with the assay of a number of hormones is that they are often present in two forms; free molecules which are physiologically active and protein-bound forms, the proteins generally being either albumin and/or special carrier proteins. Since it is the concentration of the former that is physiologically important it is necessary that the addition of labelled analyte or antibody does not affect the equilibria involved. It is possible to study some aspects of this in simple binding experiments, for instance the extent of binding of labelled $T_4$ (thyroxine) to a $T_4$-antibody should ideally not change when albumin, thyroid binding globulin or a serum sample containing these but devoid of $T_4$ is added.

### 6.5.2. Detection limits and sensitivity

The detection limits and sensitivity of an assay depend largely on two factors, the precision of the measurements involved in detecting the signal, and the slope of the signal-concentration response curve (see Part 1).

A precision study is generally carried out by assaying a sample many times both within one assay (within-assay precision), and for a number of different assays (between-assay precision). These studies can in fact be carried out in many other ways to give for example between-laboratory, between-operator, between-reagent precisions. In order to allow for differences in numerical scale of results it is common practice to express the results as percentage coefficients of variation (CV), where CV is calculated as (standard deviation of replicates/mean of replicates) $\times$ 100.

The questions of detection limits and sensitivity can be more clearly understood from Fig. 6.5d.

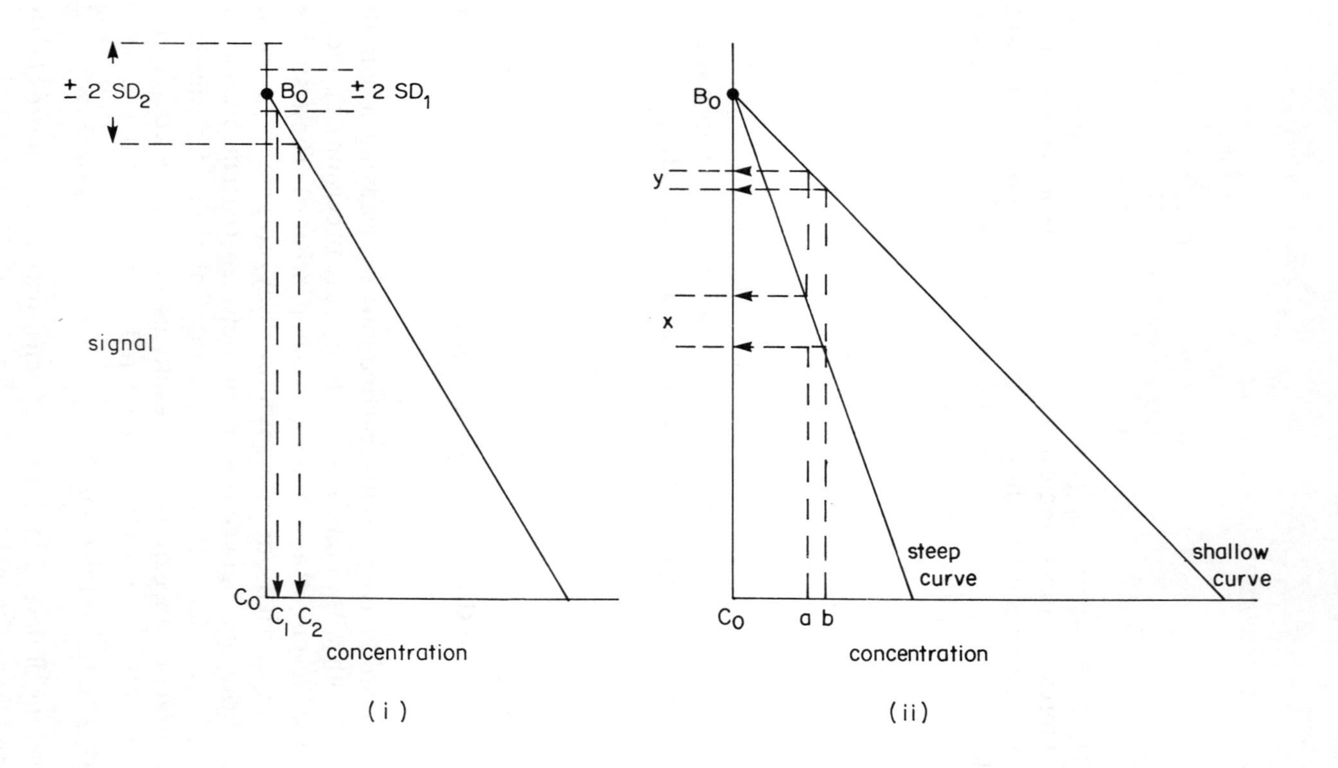


**Fig. 6.5d.** *An illustration of the concepts of sensitivity and detection limits of an assay method*

Consider the signal value $B_0$ corresponding to the binding of label in the absence of unlabelled analyte (ie at $C_0$ in Fig. 6.5d (i)). This value should be the mean of several readings and the variation of these readings may be represented by the standard deviation (SD) of the mean about $B_0$. We know that 68% of the results of replicate assays will fall within a value given by the mean $\pm 1$ SD, and 95% of the results within a range of mean $\pm 2$ SD. Hence for a precise assay the SD at $C_0$ must be small and the lowest concentration of analyte we can be confident in distinguishing from $C_0$, (95% of the time), will be smaller the greater the precision of the assay. For a situation where good precision exists at $C_0$ (ie $\pm 2\ SD_1$ is small), the lowest detectable concentration at the 95% level is $C_1$. If poor precision exists at $C_0$ ($\pm 2SD_2$ is large), the lowest detectable concentration at the 95% level is $C_2$, and you can see from Fig. 6.5d(*i*) that $C_2$ is larger than $C_1$.

For a steep standard curve the signal (x) between two similar concentrations of analyte (a and b) is greater than the corresponding signal (y) for a shallow curve (Fig. 6.5d(ii)). You will see therefore that the slope or gradient of the standard curve is important. A steep slope will enable us to distinguish more easily two concentrations since the difference in signal for the steeper standard curve is greater than for the shallower curve. However a shallow curve will operate over a wider range of concentrations. Immunoassays must therefore be optimised so that the normal range of the analyte falls in the middle of the standard curve where it is of greatest slope (remember they are in fact usually curved), and if possible this slope adjusted so that the method is able to distinguish the concentrations of likely samples over the concentration range expected.

### 6.5.3. Correlation with other methods

In spite of all of the precautions taken, there may still be a component or an unknown metabolite in the sample that will interfere in the assay, or some other reason why the method gives inaccurate or imprecise results. It is therefore advisable to compare the results obtained using any newly developed method with those obtained with the same samples in an established technique, preferably one based upon different physical or chemical principles. For example we could compare the results obtained using RIA and a technique such as HPLC, and calculate the correlation coefficient as a measure of the relationship between them.

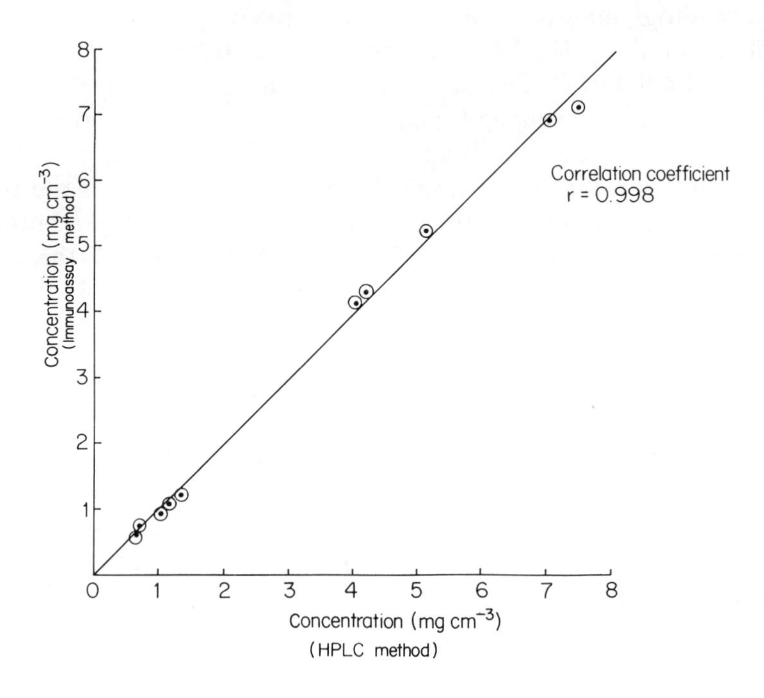


**Fig. 6.5e.** *A graphical technique for the comparison of assay values obtained using an immunoassay method and another established method*

If the results obtained by the two methods are identical, the values will be related by the equation for a straight line, $y = mx + c$ (where $y$ are the results obtained by one method, and $x$ the results obtained by the other, $m$ the gradient of the line and c the intercept on the $y$ axis).

**SAQ 6.5b**

For identical results between two assay methods, a correlation coefficient of 1 should be obtained, what should be the values of $m$ (the slope) and $c$ (the intercept)?

**SAQ 6.5b**

**SAQ 6.5c**

The drug diphenylhydantoin is of major significance in the treatment of epilepsy, particularly the 'grand mal' syndrome. Since the clinical symptoms of epilepsy are not easily monitored it is particularly important to be able to assay the serum concentrations frequently, easily and reliably. Many methods using different principles have been evolved and the following data show comparisons between ⟶

**SAQ 6.5c (cont.)**

(*i*) a radioimmunoassay and an enzyme immunoassay method, and

(*ii*) the radioimmunoassay and a gas chromatographic method.

The correlation coefficient ($r$) can be calculated from the formula

$$r = \frac{\Sigma xy - \frac{(\Sigma x)(\Sigma y)}{N}}{\sqrt{\left(\Sigma x^2 - \frac{(\Sigma x)^2}{N}\right)\left(\Sigma y^2 - \frac{(\Sigma y)^2}{N}\right)}}$$

Where $x$ and $y$ are the two sets of data and $N$ the number of data points.

Calculate the correlation coefficient and comment on the results. Are there any other simple statistical investigations that could usefully be made?

Concentration of diphenylhydantoin (mg dm$^{-3}$) by:

| Radio-immunoassay | Enzyme-immunoassay | Gas-liquid Chromatography |
|---|---|---|
| 5 | 7 | 7 |
| 7 | 9 | 10 |
| 6 | 8 | 8 |
| 8 | 11 | 12 |
| 9 | 11 | 11 |
| 5 | 6 | 7 |
| 10 | 13 | 13 |
| 6 | 8 | 12 |
| 8 | 9 | 9 |
| 9 | 12 | 12 |

**SAQ 6.5c**

## 6.5.4. Quality Control

As with all other analyses, frequent and diverse investigations to monitor the 'quality', ie general reliability of the numerical results obtained are necessary. These can take many forms and reference to other volumes in this ACOL series, particularly *The Assessment and Control of Biochemical Assays*, could usefully be made.

However there are a number of problems that arise very commonly in immunoassays that will provide an interesting closing discussion. One such problem centres around the comparison of results using kits or home-made reagents containing different antibodies.

Take the case of parathyroid hormone (PTH) which is primarily responsible for the regulation of calcium, and the assay of which commonly uses antigens directed against the carboxyl end of the protein. Such antigens measure both inactive carboxyl-terminus fragments and fully active molecules, but unfortunately the relative proportions of the two can change as a result of malfunctions in the main site of breakdown, the kidney. Serious conditions such as end stage renal disease will double the half-life of PTH, while increasing that of the carboxyl fragment about 25 fold. A marked increase in immunoreactivity becomes apparent that is not a true reflection of the physiological state. It is very likely that different antisera, particularly those directed against other parts of the molecules do not show this effect, or do so to a different degree. Comparisons then become difficult.

**SAQ 6.5d**

The experimental situation that forms the basis of this SAQ illustrates a common problem in the performance of ELISA and similar assays in which the binding of reagents to the plastic walls of microtitre plates is involved and serves to highlight the need to pay attention to the quality and reproducibility of apparatus, reagents etc in these assays. It is essential that the plastic should have consistent binding properties particularly across each plate but also ideally between plates.

The following data show the absorbance values of the contents of the wells of a typical plate of 12 columns and 8 (A–H) rows. While there are sophisticated statistical techniques for the comparison of such data can you make a subjective judgement as to whether any region of the plate is giving atypical results for the supposedly 96 replicate assays. Mean value is 1.06

| | 1 | 2 | 3 | 4 | 5 | 6 | 7 | 8 | 9 | 10 | 11 | 12 |
|---|---|---|---|---|---|---|---|---|---|---|---|---|
| A | 0.86 | 0.89 | 0.87 | 0.94 | 0.99 | 1.00 | 0.94 | 0.95 | 0.95 | 0.99 | 0.99 | 0.83 |
| B | 0.86 | 0.90 | 0.97 | 1.03 | 1.15 | 1.26 | 1.03 | 1.00 | 0.99 | 1.00 | 0.88 | 0.85 |
| C | 0.66 | 0.90 | 1.06 | 1.14 | 1.26 | 1.29 | 1.20 | 1.10 | 1.10 | 1.01 | 0.90 | 0.84 |
| D | 0.84 | 1.00 | 1.18 | 1.31 | 1.43 | 1.50 | 1.34 | 1.50 | 1.22 | 1.20 | 0.90 | 0.82 |
| E | 0.86 | 1.01 | 1.28 | 1.37 | 1.52 | 1.36 | 1.36 | 1.38 | 1.32 | 1.13 | 0.89 | 0.83 |
| F | 0.85 | 0.86 | 1.13 | 1.28 | 1.34 | 1.35 | 1.29 | 1.34 | 1.27 | 1.08 | 0.85 | 0.80 |
| G | 0.78 | 0.93 | 1.02 | 1.17 | 1.10 | 1.25 | 1.24 | 1.24 | 1.18 | 1.06 | 0.90 | 0.81 |
| H | 0.75 | 0.78 | 0.92 | 0.94 | 1.07 | 1.18 | 1.21 | 1.18 | 1.04 | 0.98 | 0.90 | 0.76 |

**SAQ 6.5d**

**Summary**

While a wide range of quantitative immunoassay techniques is now available, nearly all are based on one or other of a small set of physical principles, in particular the heterogenous (separation requiring) competitive method using a labelled ligand in competition with unlabelled ligand for a limited supply of a binding agent. The general procedure for such an assay is described in detail. Three distinct types of label (radioisotopes, fluorescent molecules, and enzymes) have been used in the majority of published methods, and they have quite distinct merits and faults. In addition a relatively large number of alternative labels have also been employed in specialised applications, and a comprehensive discussion of the labels is given.

While the basic principles of quantitative immunoassays are fairly straightforward, there are many potential problems with the assays; as a consequence a very careful study of both the immunological aspects of the method, and of purely practical considerations, needs to be undertaken. The principal aspects of such a study are described.

**Objectives**

You should now be able to:

- describe or illustrate diagramatically, the fundamental principles of a heterogeneous competitive labelled ligand assay;

- explain why a linear transformation of the curved relationship between signal and concentration is useful and how it is usually achieved in this field;

- list the properties of an ideal label for immunoassays, and give some examples of common and less common labels used in published methods;

- compare the properties of the two most common radioisotope labels and discuss their applications and limitations;

- describe some of the different ways in which fluorescent and luminescent labels have been used;

- explain the principles of the ELISA and EMIT assay methods, and state the major limitation to the use of enzyme substrates as labels;

- discuss the main separation methods currently employed in quantitative assays and indicate why it is useful to avoid the need for a separation stage if possible;

- list the principal criteria on which method selection is based;

- state some of the more important potential immunological problems with immunoassays and briefly how they can be investigated;

- describe the factors determining the sensitivity and detection limits of a method;

- state (in non-mathematical terms) what is meant by a correlation coefficient, and outline its usefulness in the evaluation of a newly developed immunoassay.

# Self Assessment Questions and Responses

**SAQ 1.2b**

Select from the following options valid reasons for using bioassays involving animals as a last resort.

(*i*) They may be expensive and time consuming to perform.

(*ii*) They can only be used for biologically-active compounds.

(*iii*) They generally give poor precision.

(*iv*) In some types of assays animals may die or have to be slaughtered.

**Response**

Option (*ii*) is a good reason for actually using a bioassay. It may well be that all alternative assay methods include measurements of inactive precursors or metabolites of the biologically-active compounds.

The other three options all provide ample justification for regarding bioassay which use animals or their tissues as 'last resort' methods.

**********************************

**SAQ 1.3a**

Indicate which of the following statements about bioassays are correct.

(*i*) Bioassays can be applied to any analyte.

(*ii*) Pure analytical grade standard is required in bioassays.

(*iii*) Bioassays measure the biological response of a living organism to the analyte.

(*iv*) The greater the biological response of the organism to the analyte the greater the potency of the analyte.

(*v*) The potency and biological activity of the analyte are synonymous.

(*vi*) The potencies of analytes in bioassays are expressed in mass units of mass per volume.

**Response**

Statement (*i*) is false; bioassays can only be used for substances that produce biological responses in living organisms and tissues. This is a major limitation in the use of these assay systems in general chemical analysis.

Statement (*ii*) is false; bioassays can be performed with pure standards but they are not prerequisites. Standards with arbitrarily assigned values are most commonly used.

Statements (*iii*), (*iv*) and (*v*) are true.

Statement (*vi*) is false; the standards used in bioassays do not have their compositions defined in terms of mass per unit volume. Instead arbitrary units of potency are assigned to the standard reference material with which test sample responses are compared.

***********************************

**SAQ 1.4a**

Which of the following could be classed as procedures that give quantal responses?

(*i*) Doses of weedkiller which cause the death of rats.

(*ii*) Doses of toxin which produce visible convulsions in the hind legs of guinea-pigs.

(*iii*) Doses of prothrombin on the time required for fibrin clots to form in plasma samples.

(*iv*) Doses of radiation on the life span of selected laboratory animals.

**Response**

The response in example (*i*) is quantal since the animals either die or they do not.

With example (*ii*) convulsions could vary in their intensity but visual observations are hardly likely to be capable of providing a reliable scaled response. In this context a convulsion either does or does not occur, ie it is a quantal response.

In example (*iii*) time is measured with a scale of possible responses. It is not therefore a quantal response.

Death in example (*iv*) serves as an indicator of the end of the life span. However it is the life span which is measured as a variable time; hence not a quantal response.

***********************************

**SAQ 1.5a**

Four groups of animals were subject to a total of four doses of drug comprising high and low standard doses (SH and SL) and high and low test doses (TH and TL). A cross-over assay was then performed using the following pattern:

| | First exposure | Second exposure |
|---|---|---|
| Group A | SH | TH |
| Group B | SL | TL |
| Group C | TH | SH |
| Group D | TL | SL |

How could randomisation of the above pattern be improved?

**Response**

Each group of animals received either a high or a low dose of the drug, whereas it might have been better to have mixed the dose more, eg SH followed by TL for group A rather than SH followed by TH etc.

***********************************

**SAQ 1.5b**

In an insulin assay at least 96 mice are needed and are randomly distributed into 4 groups (two test and two standards).

It would be possible to arrange the four groups by catching the first 24 mice and designating them as 1, the next 24 as 2 and so on. ⟶

**SAQ 1.5b (cont.)**

Can you give a fairly simple reason why this practice should *not* be adopted.

**Response**

It could be argued that the first group of 24 animals represent the less lively individuals which are thus the easiest to catch, or they may have been some other reason which caused them to congregate in small groups. It would be far better to adopt a more random method of selecting the 4 groups which avoids these problems of bias in the assay.

***********************************

**SAQ 2.1a**

The table shows numbers of *Escherichia coli* per $cm^3$ in a liquid culture medium over a period of incubation at 37 °C.

Convert the counts to logarithms to the base 10 and plot the logarithmic values against time on the graph.

Calculate the generation time for *Escherichia coli* over a period of incubation when there is a straight line relationship between the logarithm of numbers present and time.

| Time (min) | Count (no. per $cm^3$) | log (count) |
|---|---|---|
| 60 | $1.6 \times 10^2$ | |
| 120 | $2.3 \times 10^2$ | |
| 180 | $1.7 \times 10^3$ | |
| 240 | $1.0 \times 10^4$ | |
| 300 | $7.1 \times 10^4$ | |
| 360 | $4.7 \times 10^5$ | |
| 420 | $3.1 \times 10^6$ | |
| 480 | $2.0 \times 10^7$ | |
| 540 | $1.3 \times 10^8$ | |
| 600 | $2.5 \times 10^8$ | |

**Response**

| Time (min) | Count (no. per $cm^3$) | log (count) |
|---|---|---|
| 60 | $1.6 \times 10^2$ | 2.20 |
| 120 | $2.3 \times 10^2$ | 2.36 |
| 180 | $1.7 \times 10^3$ | 3.23 |
| 240 | $1.0 \times 10^4$ | 4.00 |
| 300 | $7.1 \times 10^4$ | 4.85 |
| 360 | $4.7 \times 10^5$ | 5.67 |
| 420 | $3.1 \times 10^6$ | 6.49 |
| 480 | $2.0 \times 10^7$ | 7.30 |
| 540 | $1.3 \times 10^8$ | 8.11 |
| 600 | $2.5 \times 10^8$ | 8.40 |

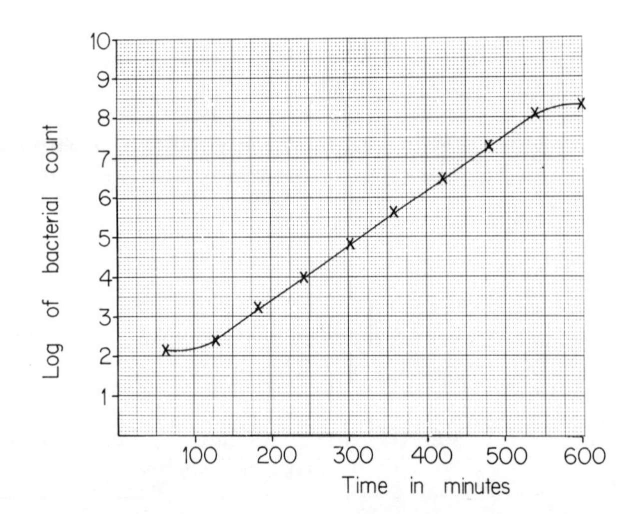


From the graph it is evident that a straight line relationship exists between 120 and 540 minutes. Taking those times as $t_1$ and $t_2$ respectively the logarithms of the bacterial counts ($b_1$ and $b_2$) are 2.36 and 8.11. Substituting these results in Eq 2.1d:

$$n = 3.322 \times (8.11 - 2.36)$$
$$n = 19.10$$

There were therefore 19.10 divisions over the incubation period of 120 to 540 minutes. 19 divisions would be a more reasonable figure to report but correction at this stage is not necessary.

Substituting the calculated value of $n$ in Eq 2.1e:

$$\text{Generation time} = (540 - 120)/19.10$$
$$= 22 \text{ minutes}$$

This figure has been corrected to the nearest whole number.

*********************************

**SAQ 2.1b**

Mark with a cross whether each of the following is associated with culture in a broth medium or a surface-inoculated agar medium. You may assume that about the same number of bacteria are inoculated and that the volumes of broth and agar media are similar.

| | Broth medium | Agar medium |
|---|---|---|
| Greater availability of nutrients | | |
| Higher yield of bacteria | | |
| Growth of bacteria as colonies (clumps) | | |
| Greater sensitivity to the atmosphere | | |

**Response**

| | Broth medium | Agar medium |
|---|---|---|
| Greater availability of nutrients | X | |
| Higher yield of bacteria | X | |
| Growth of bacteria as colonies (clumps) | | X |
| Greater sensitivity to the atmosphere | | X |

Comments:

1. The liquid growth medium allows rapid diffusion of nutrients and of bacteria by convection currents and by the motility of the latter in some cases. In an agar culture diffusion occurs less readily and bacteria are immobilised save for a few species which have the ability to swarm over the surface, eg *Proteus* species.

2. The higher yield of bacteria follows mainly from the greater availability of nutrients and removal of toxic waste products by dilution.

3. Surface inoculation of agar medium gives growth of bacteria as discrete colonies. This makes the detection of growth easy and allows the microbiologist to establish and confirm that the culture is not contaminated with unwanted microbial species.

4. Surface growth is exposed directly to the surrounding atmosphere and is therefore particularly sensitive to changes in oxygen and carbon dioxide tensions or to the humidity. Broth cultures are less sensitive.

************************************

**SAQ 2.4a**

Is the composition of the assay medium the same or different from the culture medium which is normally used to grow the test organism in the assay of:

(*i*) growth inhibiting substances;

(*ii*) growth promoting substances?

**Response**

(*i*) For growth inhibiting substances the composition of the assay medium and the normal culture medium is the same.

In the assay of growth inhibiting substances the medium should be capable of giving a good growth of the test organism. It is possible to use the same medium for assays as for normal culture as long as it does not contain any substances which might interfere with the activity of the analyte.

(*ii*) For growth promoting substances the composition of the assay medium and the normal culture medium is different.

In the assay of growth promoting substances the assay medium must not contain any of the analyte. It is important to ensure that any traces of analyte are not carried over from the normal growth medium when the test organism is inoculated into the assay medium.

*********************************

**SAQ 2.4b**

In a plate assay, are zone diameters likely to increase or decrease as a result of:

(*i*) reducing the thickness of the medium;

(*ii*) increasing the number of organisms added to the assay medium?

**Response**

(*i*) Reducing the thickness of the medium increases the zone diameter.

Zone sizes will increase as the thickness is reduced since effectively there is less medium to dilute the diffusing sample.

(*ii*) Increasing the number of organisms added to the assay medium decreases the zone diameter.

An increase in the number of organisms will tend to meet with more of the analyte giving decreased zone diameters.

***********************************

**SAQ 2.4c**

Well A has been cut properly (no splits) and a test solution diffuses radially into the surrounding medium to produce a perfectly circular zone.

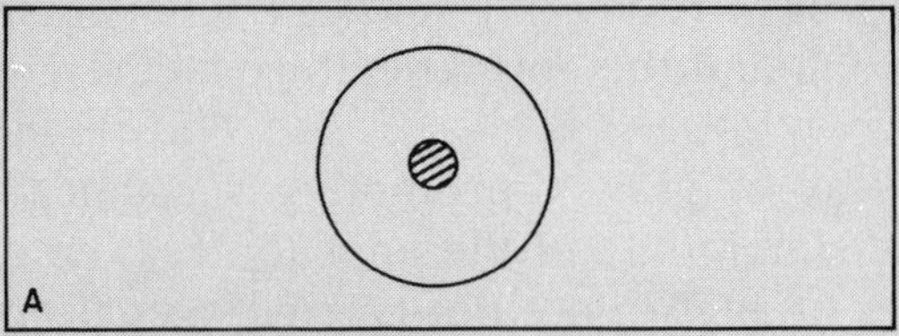

A

Well B has a split on one side. Draw the outline of a diffusion zone that might form after a test solution has been added to this well.

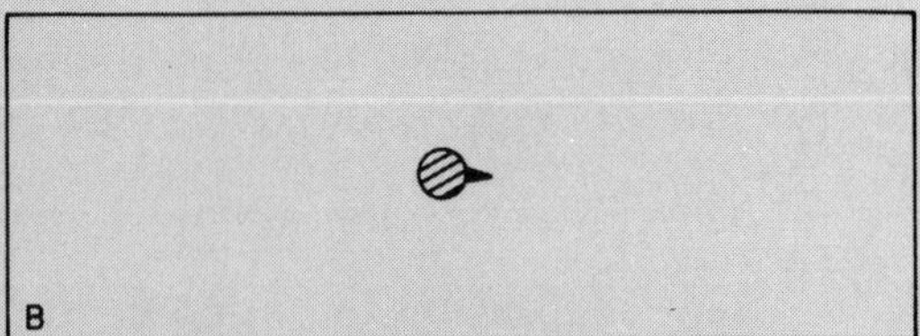

B

**Response**

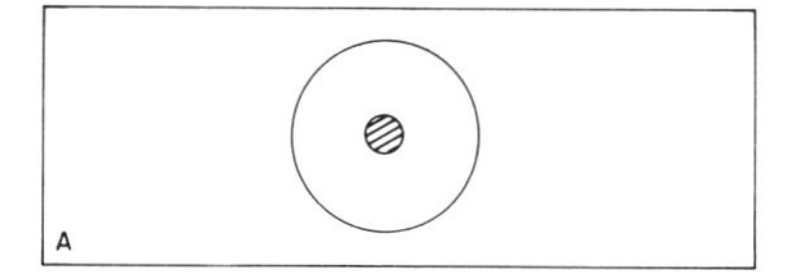


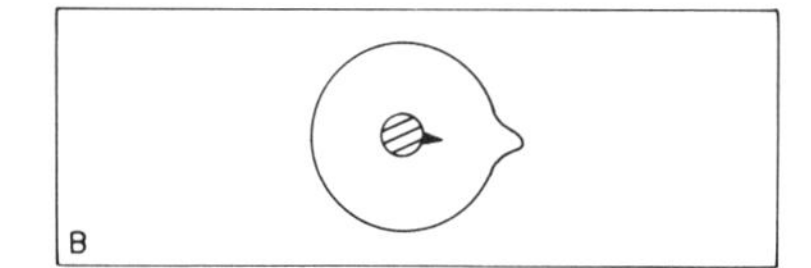


Diffusion from the split in well B produces a distorted zone. Measurements taken from such a zone have no validity.

************************************

**SAQ 2.4d**

The table shows zone diameters produced around wells filled with identical volumes of a series of gentamicin standards. Plot graphs of zone diameters versus gentamicin concentration (graph A) and zone diameters versus the logarithm (to the base 10) of gentamicin concentration (graph B)

| Zone diameter (mm) | Gentamicin ($\mu$g cm$^{-3}$) | log (gentamicin concentration) |
|---|---|---|
| 19.5 | 1 | |
| 20.8 | 2 | |
| 21.6 | 3 | |
| 22.2 | 4 | |
| 22.6 | 5 | |
| 23.0 | 6 | |

**Response**

| Zone diameter (mm) | Gentamicin ($\mu$g cm$^{-3}$) | log (gentamicin concentration) |
|---|---|---|
| 19.5 | 1 | 0.00 |
| 20.8 | 2 | 0.30 |
| 21.6 | 3 | 0.48 |
| 22.2 | 4 | 0.60 |
| 22.6 | 5 | 0.70 |
| 23.0 | 6 | 0.78 |

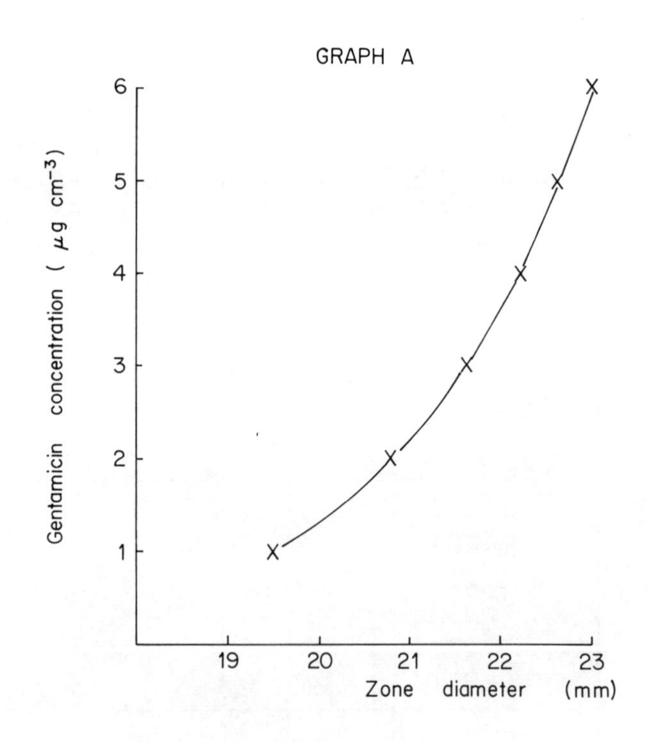

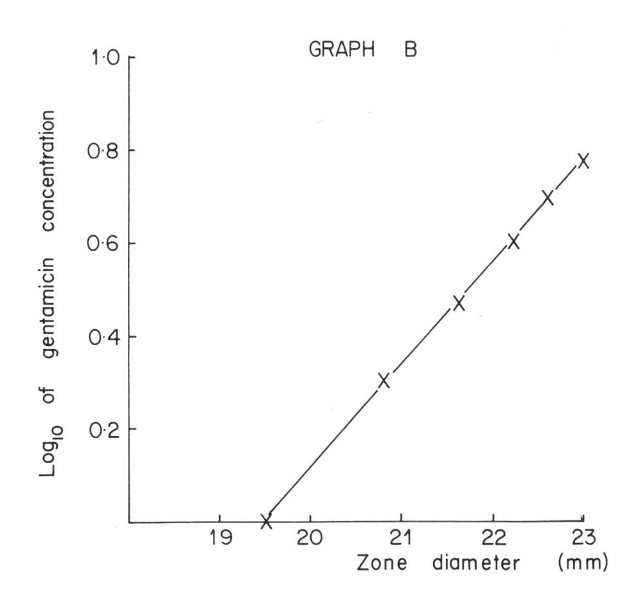


************************************

**SAQ 2.4e**

Can you explain why the optimum growth temperature is used for microbiological assays in spite of it being the temperature which gives smaller zone diameters and thus lower sensitivity. Reference to the graph showing how temperature affects the growth of a typical mesophilic bacterium may help with your explanation.

⟶

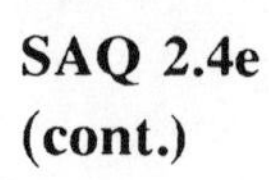

**SAQ 2.4e (cont.)**

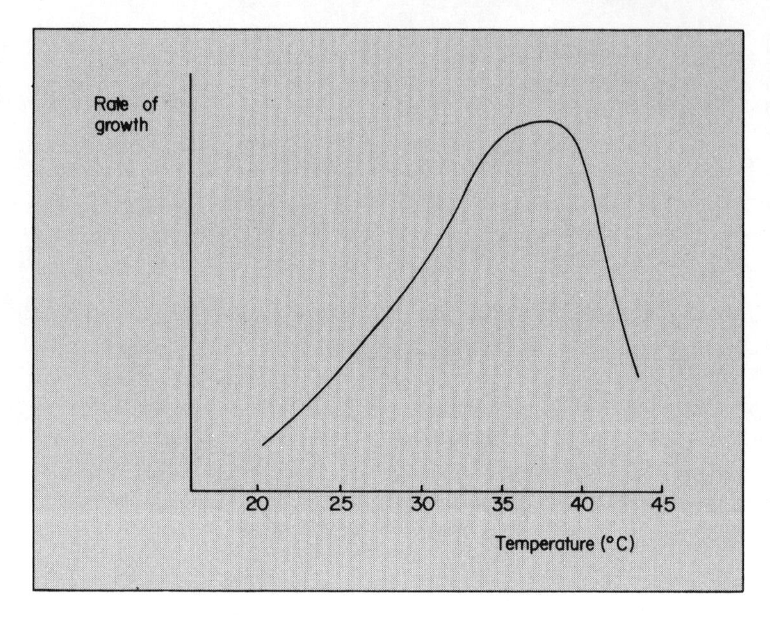


**Response**

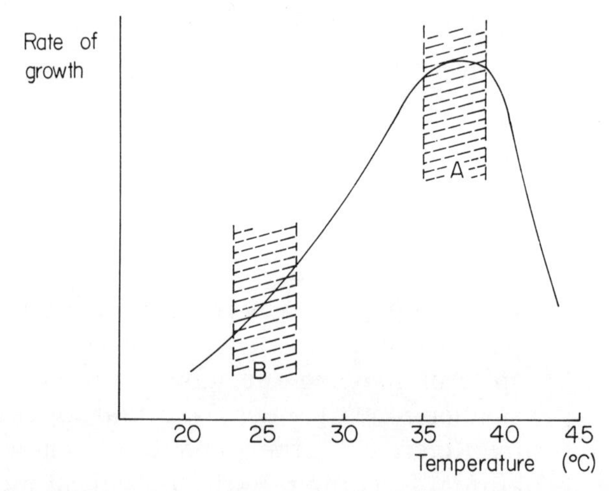


Greater reproducibility is obtained when the organism is grown at its optimum temperature.

Shaded area A on the graph shows that fluctuations in incubator temperature of 2 °C around the optimum 37 °C have little effect on growth rate. Shaded area B demonstrates considerable differences in growth rate for the same temperature fluctuations around 25 °C.

**********************************

**SAQ 2.4f**

The results of a 2 + 2 assay with *Bacillus subtilis* as the test organism and streptomycin at a ratio of 4 : 1 are tabulated.

| Treatment | Mean zone diameter (mm) |
|---|---|
| High standard (80 U $cm^{-3}$) | 20.40 |
| Low standard (20 U $cm^{-3}$) | 16.05 |
| High test sample used undiluted | 20.93 |
| Low test sample at 1/4 dilution | 16.50 |

Plot log (standard dose) against the zone diameter and draw a straight line between the two points. Plot the test sample values as if the undiluted preparation contained 80 U $cm^{-3}$ and the 1/4 dilution 20 U $cm^{-3}$ against their respective zone diameters (20.93 and 16.50 mm)

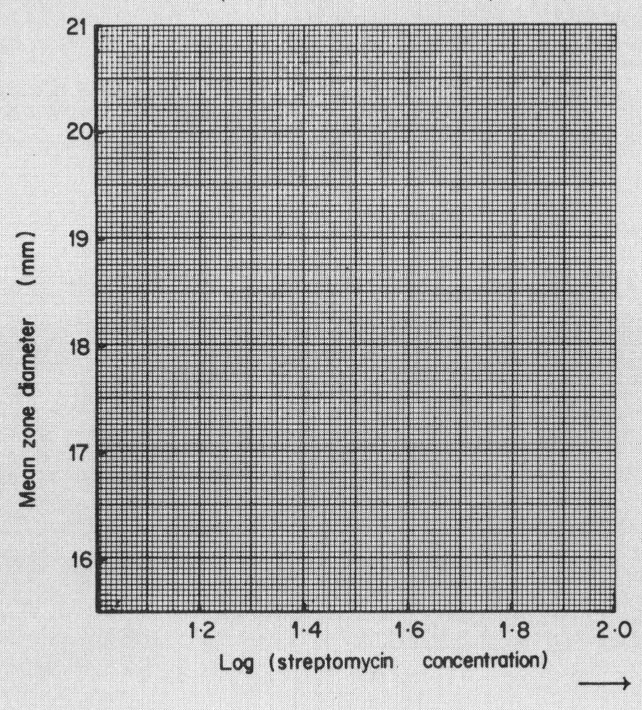

**SAQ 2.4f (cont.)**

Note that your lines should be almost parallel which indicates that the assay was valid. Lack of parallelism is associated with invalid assays, and may for example indicate that the organism is responding differently to the test and sample materials. Measure the average distance ($M$) from the test to the standard line in terms of (streptomycin dose). This value is the logarithm of the relative dose of the test to the standard. Its antilog is called the potency ratio.

$$\text{potency ratio} = \frac{\text{test dose}}{\text{standard dose}} = \text{antilog } M$$

The product of the potency ratio and the high standard dose gives the dose values of the undiluted test sample:

$$\text{Test dose} = \text{potency ratio} \times \text{high standard dose.}$$

Calculate the potency ratio and the streptomycin level of the test sample.

**Response**

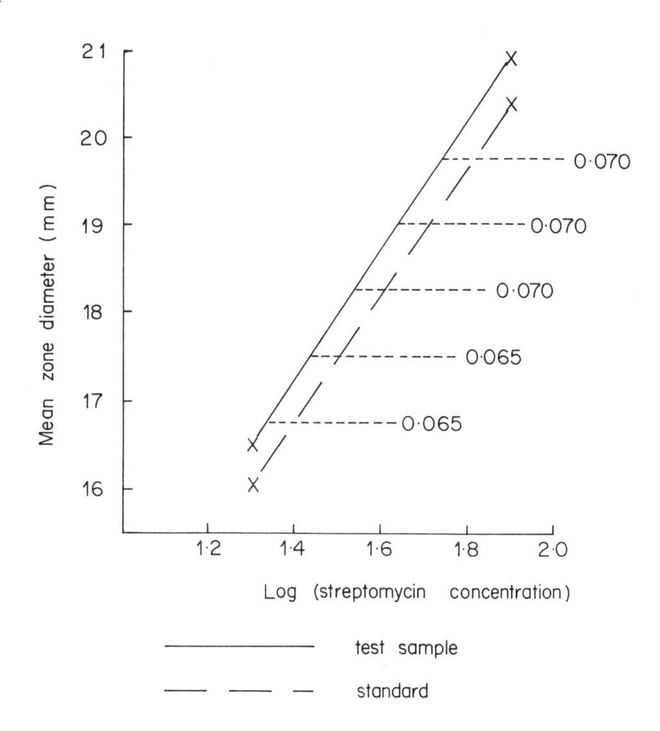


We took the mean of the differences between the response lines at the given points on the graph as follows:

0.065
0.065
0.070
0.070
<u>0.070</u>

Mean value = 0.34/5 = 0.068

The potency ratio = antilog of 0.068 = 1.1690

Concentration of streptomycin in the test sample (ie test dose) =

$$80 \times 1.1690 = 93.52 \text{ U cm}^{-3}$$

*********************************

**SAQ 2.4g**

An analyst recognised the validity of calculating results using mean values derived from a large number of replicate tests. She set out ten replicate Petri dishes for 2 + 2 assays on each of ten different test samples making one hundred dishes in all. Her sequence of pipetting out the solutions into wells was in batches as follows:

*First*. All the high standard doses, ie 10 repeated manipulations from the same bottle.

*Second*. All the low standard doses, ie 10 repeated manipulations from the same bottle.

*Third*. The high test sample doses, ie 10 repeated manipulations from each of 10 bottles (100 in all).

*Fourth*. The low test sample doses, ie 10 repeated manipulations from each of 10 bottles (100 in all)

On completion of the work all one hundred dishes were placed in an incubator without further delay and zone diameters were measured after 24 hours. Mean zone diameters were used to plot out the graph.

The analyst stated that this method of working was less likely to result in errors, such as placing the wrong solution in a well, and it also enabled the process to be completed in the shortest time. These two reasons seem perfectly acceptable, but there is a basic flaw in her approach which we would like you to identify. You may wish to consider that each pipetting sequence from a stock bottle to a well on the dish takes an average of 12 seconds and additional time must be added on for changing from one bottle to another, changing pipette tips etc.

**Response**

Allowing for a few minutes for the analyst to change from one bottle to the next and taking account of each pipetting sequence, it seems likely that she could have not have set up the complete assay in less than one hour. This means that the doses distributed when she started working (high standard doses) had far longer preincubation times than most of the test doses (particularly the low doses).

When working with large numbers of treatments it is important to plan an assay in such a way that differences in parameters such as preincubation periods are taken into account.

**********************************

**SAQ 3.1a**

Define the terms

(*i*) antigen;
(*ii*) antibody;
(*iii*) response;
(*iv*) immune memory.

**Response**

(*i*) An antigen is a material that can by itself induce the production of antibodies.

(*ii*) Antibody is a protein synthesised in response to the presence of an antigen and which is specific for, and will bind to, that antigen.

(*iii*) The immune response is a collective name for the changes induced in the immune system by various stimuli, including antigen. These changes can be biochemical, cellular and physiological and include the production of antibodies as well as other events.

(*iv*) The immune memory is the ability of the immune system to store the information obtained from an initial encounter with an immune stimulus (eg an antigen) in order to produce a rapid and substantial response on subsequent encounters.

*********************************

**SAQ 3.1b**

Complete the following sentences with the numbers of words indicated by the lines.

(*i*) When antibody and antigen bind, they form an ____________ ____________.

(*ii*) Antigens and antibodies combine with a high degree of ____________.

(*iii*) The purpose of antibodies is to help protect against ____________ diseases.

(*iv*) The quantity of antibody produced following a second exposure to an antigen is ____________ compared with that produced on first contact.

(*v*) Antibody can, under suitable conditions, be produced to antigenic substances of experimental and commercial interest by injecting them into an animal, a process called ____________.

**Response**

(*i*) immune complex
(*ii*) specificity
(*iii*) infectious
(*iv*) large (or augmented)
(*v*) immunisation

*********************************

**SAQ 3.1c**

What do you think the term anti-insulin antibody means?

**Response**

Anti-insulin antibody is an antibody produced in response to an injection of insulin and therefore capable of binding to it.

************************************

**SAQ 3.1d**

Answer *true* or *false* to each of the following:

(*i*) the potential to produce a great diversity of antibodies is inherent in the immune system; T / F

(*ii*) the production of a particular antibody is the result of exposure to an antigen; T / F

(*iii*) the diversity of antibodies which can be produced benefits the animal or person because antibodies can be produced against a wide range of bacteria and viruses, and therefore helps resist infectious disease; T / F

(*iv*) the potential to produce antibodies of a particular specificity is the result of exposure to an antigen. T / F

**Response**

(*i*) True, the *potential* to produce a diversity of antibodies is inherent in the genetic and biochemical make-up of an organism.

(*ii*) True, the above potential will only be manifested when the synthesis of a specific antibody is induced by exposure to the appropriate antigen.

(*iii*) True, materials in, on or produced by, bacteria, viruses etc can act as antigens and stimulate the production of specific antibodies. These antibodies react with the bacteria and induce processes leading to their death. This is therefore a very important means of defence.

(*iv*) True, this is implied in (*i*)–(*iii*) above.

**********************************

**SAQ 3.2a**

From the following list select those materials that you would imagine to be good antigens.

(*i*) the enzyme RNAase with 124 amino acids of 17 different types,

(*ii*) starch – a large polymer of glucose with an $M_r$ value of $10^4$ to $10^8$,

(*iii*) glucose,

(*iv*) polypropylene,

(*v*) the large structural protein collagen, (consisting predominantly of the amino acids, proline, hydroxyproline and glycine),

(*vi*) the peptidoglycan (peptide and carbohydrate) lattice of intact bacterial cell walls.

**Response**

The following would probably be *good* antigens for the reasons indicated:

(*i*) A large and structurally complex, non-repetitive polymer.

(*vi*) A very large molecular complex, while made of repeating units, these are themselves complex.

The following would probably *not* be good antigens.

(*ii*) and (*v*) Large but structurally simple molecules.

(*iii*) A small molecule.

(*iv*) A large molecule, structurally simple and non-degradable.

**********************************

**SAQ 3.2b**

Which of the following treatments should produce a good immunological response:

(*i*) amylase from a rabbit isolated, purified and re-injected into that rabbit,

(*ii*) amylase from a rabbit isolated, purified and injected into another rabbit,

(*iii*) amylase from a rabbit injected into a horse?

**Response**

In addition to the general chemical requirements of large size, diverse structures, degradability etc for antigenic stimulation, a compound must be present in a foreign (ie 'non-self') body. This does not necessarily imply another species, simply another individual, provided they are not genetically very similar. Thus (*ii*) and especially (*iii*) should provide a good immunological response.

**********************************

**SAQ 3.2c**

(*i*) Which of the following can by themselves induce antibody formation:

antigens,
haptens,
immunogens?

(*ii*) What do the following terms mean:

epitope,
antigenic determinant,
valency?

(*iii*) How many binding sites would you expect to find on:

antigens,
haptens?

(*iv*) Which of the following would have a greater diversity of binding sites:

carbohydrates
proteins?

(*v*) Give an alternative name for antibodies.

(*vi*) Distinguish between monoclonal and polyclonal antibodies.

(*vii*) Is it true that the amino acid sequence in the variable region of the antibody polypeptide chain is responsible for its antigenic binding specificity? ⟶

**SAQ 3.2c (cont.)**

(*viii*) Are the antibody binding sites on a single protein antigenic molecule identical?

(*ix*) When an IgG molecule is itself acting as an antigen, would all the determinants on its surface be different?

(*x*) In very simple terms how could antibody capable of binding say, anti-human insulin antibody which has been raised in guinea-pig, be prepared?

(*xi*) Suggest possible bases for the fact that antiserum raised against unpurified chicken ovalbumin (egg white protein) binds both duck ovalbumin (the corresponding protein in a different species), and chicken serum albumin (a different protein from the same species).

**Response**

(*i*) Antigens and immunogens both induce antibody formation. The term immunogen is also used in a wider sense to mean a substance inducing any immunological response. Haptens are low molecular mass compounds and need coupling to another molecule to increase the effective molecular mass to one which will induce antibody formation.

(*ii*) Epitope and antigenic determinant both mean an individual binding site on the surface of an antigen. Valency is the number of such binding sites per molecule.

(*iii*) Antigenic molecules usually have many binding sites due to their size, and certainly should have many to be effective as antigens. Haptens usually have only one.

(*iv*) Proteins, since most contain a reasonable number of different amino acids as their monomers whereas many carbohydrates have only one or two sugars as their monomeric units.

(*v*) An alternative name for antibodies is immunoglobulins.

(*vi*) Monoclonal antibodies are pure antibodies, ie all molecules are of exactly the same structure. Polyclonal antibodies are mixtures of structurally different molecules although they may all react with the same antigen.

(*vii*) This is true, it is the precise amino sequence in this region that affects the shape, the physical properties (eg charge distribution) and hence, reactivity of the molecule and gives rise to its specificity. Variation in the constant region distinguishes the different classes of antibodies (IgG,IgM etc).

(*viii*) No, proteins consist of chains of up to about twenty different amino acids usually in a random, although critical, sequence. The residues in any given locality are likely to form a fairly unique pattern and hence each antigenic determinant site will almost certainly be different.

(*ix*) No, as with other proteins there will be many different determinants on the surface, but because of the bilateral symmetry of the IgG molecule, all sites will be duplicated.

(*x*) This antibody would usually be obtained by immunising an animal of another species, eg a goat or rabbit, with antibody from the guinea-pig (giving an anti-guinea-pig antibody). If the guinea-pig had in turn been immunised with human insulin, its antibody should be capable of binding the insulin, ie would be anti-human insulin antibody.

(*xi*) Corresponding proteins in different species are often structurally very similar and the cross-reaction is probably due to the duck protein having some determinants in common with and/or similar to those on the chicken ovalbumin. It is less likely that chicken serum albumin has many binding sites in common with the ovalbumin, although as they are both albumins, this is possible. More probably the original ovalbumin immunogen was contaminated with some serum albumin.

**********************************

**SAQ 4.2a**

(*i*) Which of the following pairs associate by covalent and which by non-covalent bonds:

antigens and antibodies;
haptens and carrier molecules?

(*ii*) Which of the following should be regarded as a hapten rather than as an antibody:

egg white albumin;
haemoglobin;
thyroxine (a dipeptide hormone)?

(*iii*) About what proportion of the total antibody would tend to be bound, in each case, if a large quantity of hapten was mixed with a small (ie limiting) quantity of the appropriate:

monoclonal antibody;
polyclonal antibody?

(*iv*) Give two of the limitations to the use of antibodies as reagents.

**Response**

(*i*) Haptens and carriers link by covalent bonds, whereas antigens and antibodies associate using non-covalent forces.

(*ii*) Thyroxine has an $M_r$ value of only 776 and is therefore a poor antigen. It is best regarded as a hapten.

(*iii*) A high proportion (perhaps virtually all) of the specific monoclonal antibody should bind to the hapten, whereas a lower proportion of the mixed polyclonal type would be bound.

(*iv*) Among the more important limitations are:

- the possible lack of absolute specificity of inter-action;
- the fact that an assay method using antibodies relies upon an immunological reaction, and this may not necessarily reflect the physiological or biochemical activity of the material being assayed;
- the antibody-antigen reactions may be ill-defined and poorly reproducible, especially if impure samples and antibody preparations are used;
- while antibodies are in fact relatively stable, they are nonetheless proteins and can be adversely affected by the environment;
- production of specific antibodies can be expensive and demand considerable technical skill.

**************************************

**SAQ 4.3a**

(*i*) Describe two ways in which complete adjuvant can increase antibody formation.

(*ii*) Name three constituents of a complete adjuvant.

(*iii*) What weight of antigen is usually regarded as maximal for a given series of injections in a small laboratory animal?

(*iv*) What are the usual routes of injection during the production of antibody?

(*v*) Which of the following statements about the quantity of antibody produced in an immunisation programme is/are correct?

⟶

**SAQ 4.3a (cont.)**

- It is directly related to the mass of dose.
- It is influenced by the site of injection.
- It varies with the molecular size of the immunogen.

(*vi*) What is serum, and what does the term antiserum mean?

**Response**

(*i*) The purpose of mixing the antigen with adjuvant is partly to ensure a slow release of the antigen from the injected site, in order to minimise any toxic effects and to prolong exposure of the immune system to the stimulus. The complete adjuvant should also stimulate the immune system, see (*ii*).

(*ii*) Complete adjuvant contains oil and detergent to form a stable emulsion, and also a stimulant of the immune system.

(*iii*) For a laboratory animal (rabbit, mouse etc) doses of less than 1 mg are usually given.

(*iv*) Nowadays injections are usually either sub-cutaneous and/or intra-muscular.

(*v*) False, the quantity of antibody produced is not directly related to the dose, although a high dose can inhibit (paralyse) the immune system.

True, the quantity of antibody produced is influenced by the site of infection, although this is not usually a problem.

True, the quantity of antibody produced varies with the molecular size of the immunogens, this aspect has been discussed previously in the context of antigens and haptens.

(*vi*) Serum is the clear fluid left after whole blood clots and the clot has been removed. It contains many of the original blood proteins (including any antibodies) and virtually all of the low molecular mass components. Antiserum is serum containing antibodies.

***********************************

**SAQ 4.4a**

If a conjugate of the steroid hormone progesterone and a carrier is used to immunise a mouse and the lymphocytes removed, fused to form hybrids which are grown into clones, will the antibodies produced by the different clones be identical?

**Response**

No, a number of different lymphocytes may respond to the progesterone/carrier conjugate, and even to the same site on the conjugate, but they will respond in different ways. The monoclonal antibodies produced by the clones descending from each of these cells may differ in strength of binding, degree of specificity etc, although they should all respond to the progesterone hapten.

***********************************

**SAQ 4.4b**

Answer each of the following questions with the term 'monoclonal' or 'polyclonal'.

(*i*) Which type of antibody requires the more complex initial preparation and is in general more difficult to produce?

(*ii*) For the production of which antibody is the purity of the initial immunogen of greater importance?

(*iii*) Which antibody has less defined and more variable composition? ⟶

**SAQ 4.4b (cont.)**

(*iv*) Which antibody is not absolutely specific to the immunogen?

(*v*) Sustained large scale production is possible for which antibody?

(*vi*) Which antibody is more suitable for precipitation reactions?

**Response**

(*i*) monoclonal
(*ii*) polyclonal
(*iii*) polyclonal
(*iv*) polyclonal
(*v*) monoclonal
(*vi*) polyclonal

************************************

**SAQ 4.4c**

What is unusual about the protein metabolism of cells involved in the disease, myelomatosis, and in what way does the cell division and survival characteristics of these cells differ from those of lymphocytes?

**Response**

In myelomatosis cells produce antibody without immune stimulus, furthermore the cells multiply rapidly by uncontrolled division and the cells are very long-lived.

************************************

**SAQ 4.5a**

Which of the following statements are true?

(*i*) Both biological assays and receptor assays utilise binding by cell receptors.

(*ii*) Binding by cell receptors and serum transport proteins is generally about as specific as that by antibodies.

(*iii*) Specific serum transport proteins are available for the analysis of relatively few substances.

(*iv*) The presence of the same transport protein in the blood sample for analysis as is used as the binding reagent can complicate the analysis.

**Response**

(*i*) True, in bioassays the binding induces a physiological response in the isolated cells and tissues of the whole organism used, whereas in receptor assays, isolated receptors are used either as a binding site for the assay of the corresponding ligand, or the ligand-receptor binding can be used to assay the latter.

(*ii*) This is false, in some cases because the components in question are intentionally unspecific and multi-purpose.

(*iii*) Unfortunately this is true. Most materials of interest to analysts are not purposefully carried in the body.

(*iv*) True, since many methods require a fixed amount of binding agent, the presence of a variable amount of endogenous binder can produce quantitatively false results.

***********************************

**SAQ 4.5b**

Define the terms:

(*i*) serum transport protein;
(*ii*) receptor;
(*iii*) agonist;

**Response**

(*i*) A serum transport protein is a protein used to transport a material around the body either to control its movement or concentration, or to render it soluble in the aqueous blood system.

(*ii*) A receptor is a complex structure present on a cell surface designed to receive and respond in specific ways to particular materials eg hormones.

(*iii*) An agonist is a material to which the receptor responds.

***********************************

**SAQ 4.5c**

In the binding reaction between each of the following pairs of reactants, indicate which component would conventionally be regarded as the ligand.

(*i*) triiodothyronine/anti-triiodothyronine antibody;

(*ii*) guinea-pig anti-human insulin antibody/ human insulin;

(*iii*) guinea-pig antibody/goat anti-guinea-pig Ig antibody;

(*iv*) thyroxine-binding globulin/thyroxine.

**Response**

(*i*) triiodothyronine;
(*ii*) human insulin;
(*iii*) guinea-pig antibody (acting as an antigen);
(*iv*) thyroxine.

*********************************

**SAQ 5.1a**

List four methods by which antigen/antibody reactions can be detected.

**Response**

(*i*) Uptake of a labelled component (ie antigen or antibody) into the immune complex.

(*ii*) Precipitation of the complex, measured by eye or by following light scattering.

(*iii*) Agglutination if either the antigen or antibody is particulate.

(*iv*) Complement fixation if the antigen is on a cell surface.

*********************************

**SAQ 5.3a**

Simple labelled ligand binding assays require three components:

- a sample for analysis or a standard ligand solution,
- labelled ligand,
- the appropriate binding agent.

(*i*) Which of the above is/are present at a fixed concentration?

(*ii*) Which need(s) to be present at low concentration?

(*iii*) Using the symbols L, L* and B, what are the four main components of the equilibrium mixture?

(*iv*) As the concentration of L rises, will the bound:free ratio of label rise or fall?

**Response**

(*i*) Labelled ligand and binding agent

(*ii*) The binding agent

(*iii*) L, L-B, L*, and L*-B.

(*iv*) The ratio will fall, (more L means more L-B, hence less L*-B and a lower bound : free ratio).

**********************************

**SAQ 5.3b**

Indicate, with reasons, whether each of the following is true or false.

(*i*) In competitive binding assays the reaction is normally allowed to go to equilibrium.

(*ii*) In a simple system using a radioactive label it is necessary to separate the bound and free forms of the label in order to carry out the measurement. You need to consider (and perhaps therefore revise) the basic properties of radioisotopes and the principles of their measurement.

(*iii*) In binding assays the concentration of analyte in the samples can be determined by calculation from the known affinities of the binding agent by applying the Law of Mass Action.

**Response**

(*i*) This is normally but not absolutely necessarily the case.

(*ii*) True, since the radioactivity will be manifested equally whether the labelled component is bound or free.

(*iii*) While this is in theory true, in practice the reactions and their rates are complex, and the affinities of the various components for each other unlikely to be known accurately. The concentrations of test samples are generally determined by comparison with standards.

**********************************

**SAQ 5.3c**

By way of revision of material discussed earlier in this Unit list three reasons why antibodies are particularly useful as binding reagents.

**Response**

Among the more important are:

- specificity (generally better than other binding agents);
- affinity (generally better than other binding agents);
- stability (relatively good for a biological material);
- ease of preparation (usually more easily produced than other binding agents, especially cell receptors);
- versatility (ie antibodies can be produced to almost any antigen or hapten, and while these are usually whole molecules they can be parts of complex molecules or structures giving considerable versatility to analytical methods and qualitative studies using these techniques).

**********************************

**SAQ 5.5a**

Why is it necessary to carry out a separation of bound and free forms of the tracer before measurement when radioisotopes are used as the label?

**Response**

The emission from radioactive materials is unaffected by the physical or chemical environment of the atoms in question and hence will be given equally well by bound and free forms of the isotope. Separation will be necessary to study changes in the proportion present in one of the forms in response to changes in concentration of the test substances.

*********************************

**SAQ 5.5b**

Consider a non-competitive two-site labelled antibody assay.

(*i*) Which component is bound to the solid phase?

(*ii*) What proportion of sample antigen is bound to the first antibody?

(*iii*) How is the separation of sample antigen from the others achieved?

(*iv*) Which materials must be present in excess?

(*v*) Illustrate why this technique is sometimes referred to as a sandwich technique.

**Response**

(*i*) The first antibody

(*ii*) Virtually 100%

(*iii*) Decantation since the other antigens will not be bound to the specific antibody used

(*iv*) Both antibodies

(*v*) Support $-Ab_1-Ag-Ab_2^*$

***********************************

**SAQ 5.5c**

(*i*) Complete the following sentence with either the term competitive or non-competitive.

All labelled ligand assays for the estimation of ligand are ____________whereas most labelled antibody methods are ____________

(*ii*) Complete the following sentence with either the term, limited or excess.

Competitive assays perform optimally with ____________ reagent, whereas non-competitive assays are optimised with ____________ reagent.

**Response**

(*i*) Competitive, Non-competitive

(*ii*) Limited, Excess.

***********************************

**SAQ 5.6a**

You are required to assay the anti-epileptic drugs phenytoin, valproic acid, and carbamazepine in the blood of epileptic infants. These drugs are commonly co-administered and the range of their serum levels is 10–100 $\mu$g cm$^{-3}$ (10–100 mg dm$^{-3}$). Based upon the information given previously and your general knowledge of the design of analytical experiments, explain, with reasons, which assay system you would choose from those listed below.

(*i*) Individual immunoassays for each drug requiring 10 $\mu$l of sample per assay with each assay being automated and taking 5 minutes to perform.

(*ii*) One single HPLC assay which co-determines each drug in one run. A sample volume of 50 $\mu$l and 30 minutes assay time are required.

(*iii*) Chemical derivatisation procedures leading to spectrophotometric analyses with 10 cm$^3$ original sample and 45 minutes of time required for extraction, reaction and measurement.

**Response**

Your answer will depend upon a number of factors including:

- the instrumentation at your disposal; the availability of an automated immunoassay, HPLC or chemical analyser would obviously be a powerful incentive to adopt the appropriate method.

- the number of samples to be analysed daily; a large number making (*i*) and (*ii*) more convenient since although (*ii*) requires an assay time of 30 minutes, all 3 compounds will be measured in each run. (*i*) takes only 15 minutes for the analysis.

- the sample volume available, and for infants this is likely to be low which would make method (*iii*) unacceptable.

- the required detection limits, sensitivity, specificity etc which will depend on the exact methods involved.

- the cost-effectiveness which might make (*iii*) the only acceptable method despite its requirement for a high sample volume and a long assay time.

*********************************

**SAQ 6.1a**

Some typical results from a competitive labelled ligand radioimmunassay are tabulated.

| Standard $\mu g\ cm^{-3}$ | Radioactive counts in the precipitate, duplicate results, bound fraction (in disintegrations $s^{-1}$) | |
|---|---|---|
| 1 | 62 011 | 61 980 |
| 3 | 46 082 | 45 927 |
| 6 | 30 615 | 30 950 |
| 8 | 24 959 | 24 659 |
| 10 | 20 981 | 21 596 |
| Test sample | 36 542 | 37 218 |

Plot the results of the standards on the graph paper below using the vertical axis for the radioactive counts. Draw the best curve you can, using a 'flexicurve' if one is available. Determine the concentration in the test serum by interpolation from the graph.

**Response**

The mean values of the above data are

| Sample | | Mean Count |
|---|---|---|
| Standard | 1 | 61 996 |
| | 3 | 46 005 |
| | 6 | 30 783 |
| | 8 | 24 809 |
| | 10 | 21 289 |
| Test | | 36 880 |

When plotted the results should produce a shallow curve and a concentration for the test sample of about 4.6 $\mu$g $cm^{-3}$ (Fig. 6.1b).

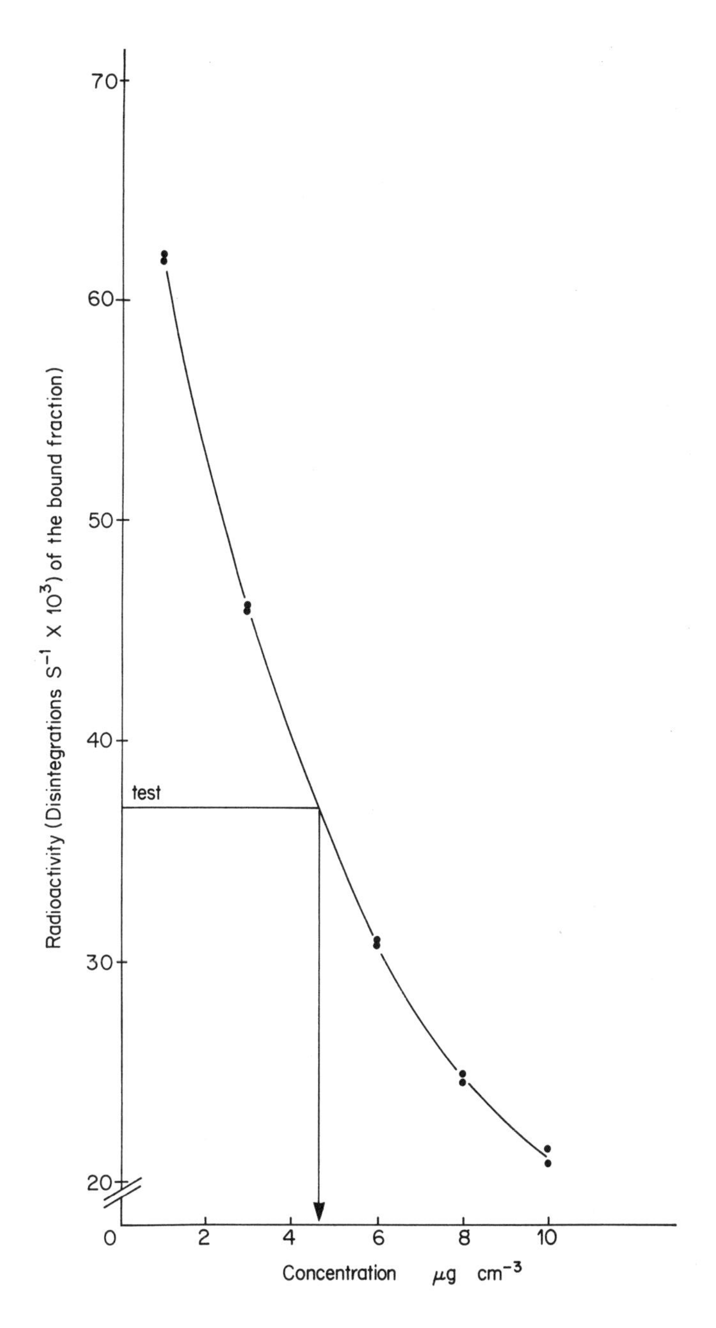


**Fig. 6.1b.** *Standard curve for a typical radioimmunoassay*

***********************************

**SAQ 6.1b**

Replot the data provided in SAQ 6.1a, given that the total radioactivity was 75 000 disintegrations $s^{-1}$, using the logit $Y$/log $X$ plot described above. To help you in this, note that the mean radioactivity of the 1 $\mu g\ cm^{-3}$ standard is 61 996 and hence the proportion bound is 61 996/75 000 or 0.827. Hence determine the concentration of the test sample.

**Response**

You should have carried out the following calculations

| $X$ | Log $X$ | Bound Label d $s^{-1}$ | Bound Proportion ($Y$) | 1-$Y$ | $Y$/1-$Y$ | ln($Y$/1-$Y$) |
|---|---|---|---|---|---|---|
| 1 | 0 | 61 996 | 0.827 | 0.173 | 4.780 | +1.57 |
| 3 | 0.48 | 46 005 | 0.613 | 0.387 | 1.584 | +0.46 |
| 6 | 0.78 | 30 783 | 0.410 | 0.590 | 0.695 | −0.36 |
| 8 | 0.90 | 24 809 | 0.330 | 0.669 | 0.493 | −0.71 |
| 10 | 1.00 | 21 289 | 0.284 | 0.716 | 0.397 | −0.93 |
| Test | | 36 800 | 0.492 | 0.508 | 0.968 | −0.032 |

A plot of these data gives a reasonable straight line

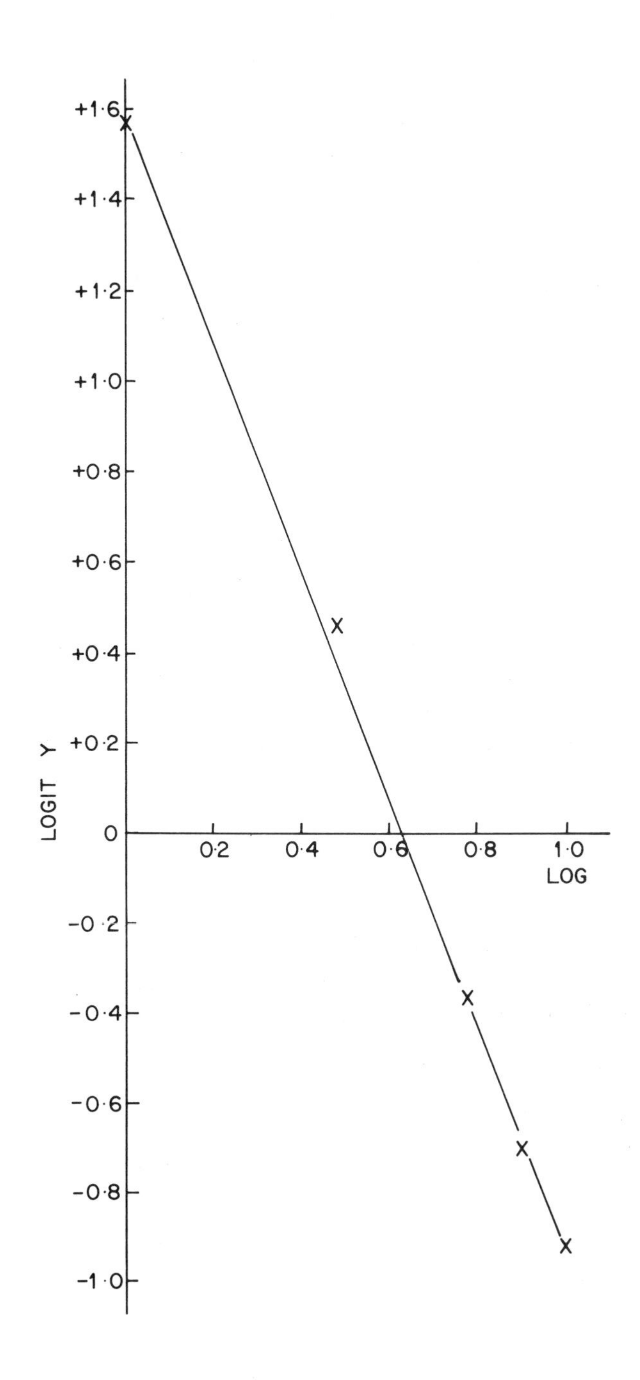
+1·6
+1·4
+1·2
+1·0
+0·8
+0·6
+0·4
+0·2
0
−0·2
−0·4
−0·6
−0·8
−1·0
LOGIT Y
0·2
0·4
0·6
0·8
1·0
LOG

From the graph the value of log(test) when ln $(Y/1-Y)$ = -0.032 is equal to 0.635. Hence concentration of test = 4.31 $\mu$g $cm^{-3}$.

***********************************

**SAQ 6.2a**

For an assay using a labelled ligand explain why $^{125}I$ will produce a method with lower detection limits than one with $^{3}H$.

**Response**

The lowest detection limit that can be achieved is dependent, among other things, upon the ability to detect and differentiate reliably, a test signal compared with the background signal. In general, the lower the latter the smaller the amount of label required to obtain a test-signal that is significantly different from it, and one advantage of using radioisotopes is that the background radiation level is very low. Since our ability to detect small amounts of radiation is very good, the detection limits are extremely low for all radioisotopes.

It is the case however, that one atom of $^{125}I$ will emit six times the radiation of one atom of $^{3}H$, and hence only 1/6 the number of $^{125}I$-labelled molecules is required to generate a given signal compared with $^{3}H$-labelled molecules. There is therefore a correspondingly lower minimum detectable concentration for the $^{125}I$ label.

***********************************

**SAQ 6.2b**

The abbreviated experimental protocols and data for two different DELFIA procedures are given.

(*i*) The assay of human luteinising hormone (LH).

- Pipette standards, experimental samples and europium-labelled anti-LH-antibody into tubes which are pre-coated with excess anti-LH.
- Incubate for 2 hours at room temperature.
- Wash off surplus reagents.
- Add the detergent solution.
- Measure the fluorescence.

| Concentration of standard. IU $dm^{-3}$ | Fluorescence c $s^{-1}$ |
|---|---|
| 1 | $2.8 \times 10^3$ |
| 5 | $1.5 \times 10^4$ |
| 50 | $1.7 \times 10^5$ |
| 250 | $9.1 \times 10^5$ |
| Sample | $9.0 \times 10^3$ |

(*ii*) The assay of cortisol.

- Pipette standards, experimental samples and europium-labelled anti-cortisol-antibody into tubes which are pre-coated with immobilised cortisol.
- Incubate for 2 hours at room temperature.

⟶

**SAQ 6.2b (cont.)**

- Wash off surplus reagents.
- Add to detergent solution.
- Measure the fluorescence.

| Concentration of standard nmol $dm^{-3}$ | Fluorescence c $s^{-1}$ |
|---|---|
| 100 | $7.0 \times 10^5$ |
| 250 | $4.8 \times 10^5$ |
| 500 | $3.5 \times 10^5$ |
| 1000 | $2.5 \times 10^5$ |
| 2000 | $1.8 \times 10^5$ |
| Sample | $4.0 \times 10^5$ |

- On the graph paper below, plot the data in each case and determine the concentration of LH or cortisol in the experimental samples. Plot sample concentration on the horizontal log axis.

- Explain the principles involved in these two procedures both of which are variants of the basic DELFIA system.

- When a sample containing 5 IU $dm^{-3}$ of LH and 5 IU $dm^{-3}$ of HCG (another of the gonadotrophin hormones) was assayed by method (*i*) the signal rose to $1.54 \times 10^4$ c $s^{-1}$. When a sample of 250 nmol $dm^{-3}$ cortisol and 250 nmol $dm^{-3}$ of corticosterone (which differs in lacking a 17-hydroxyl group) was assayed by method (*ii*) the signal obtained was $3 \times 10^5$ c$s^{-1}$. Explain this difference by considering the principles of the methods involved; you may find reference to the information in Section 5.5.1 useful.

**Response**

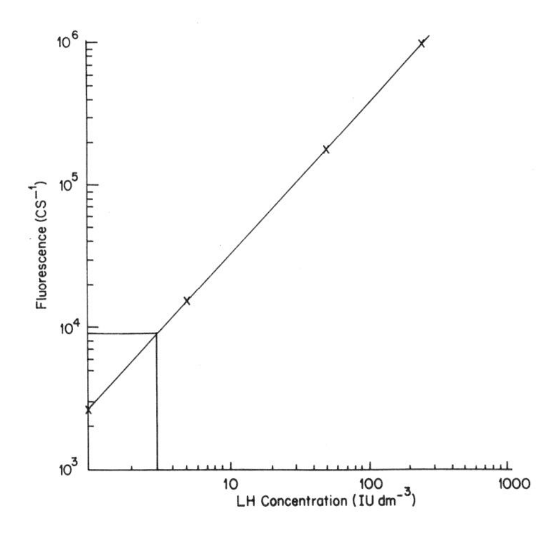


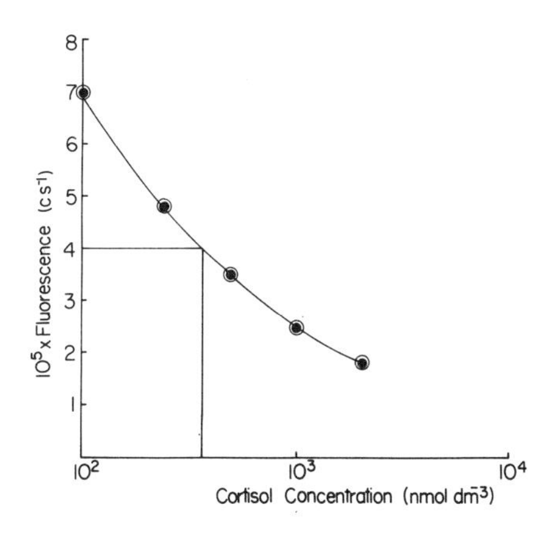


When plotted the results should appear as shown on the graphs and you should obtain values for the experimental samples of

(*i*) 3 IU dm$^{-3}$
(*ii*) 370 nmol dm$^{-3}$

In the case of method (*i*) a two site sandwich technique has been employed yielding a final product of

$$\text{Support} - \text{Ab}_{\text{LH}} - \text{LH} - \text{Ab}_{\text{LH}} - \text{Eu}$$

When detergent is added the europium chelate is released and its fluorescence measured. Remember that the act of release and formation of detergent coated micelles gives a very substantial increase in fluorescence. In this method the greater the concentration of LH the more of the above product will form and the greater the final fluorescence.

For method (*ii*) a competitive system was used with a competition between the cortisol bound to a solid phase and the soluble cortisol in the samples for a limited amount of the labelled antibody. The final product is

$$\text{Support -cortisol} - \text{Ab}_{\text{cortisol}} - \text{Eu}$$

The greater the sample cortisol concentration the less labelled anti-cortisol will bind to the solid phase and the lower the resulting fluorescence.

The results of the assays of samples containing molecules related to the target analyte indicate the specificity of the methods. Method (*i*) shows a 2.7% change and is therefore more specific than method (*ii*) which shows a 37.5% change. While this difference could be related to the degree of similarity between target analyte and the contaminating analogue in each case, it is also probable that the use in (*i*) of a double antibody procedure where the antibodies would be directed against different determinants on the LH molecule would be a major contributor to the specificity obtained.

***********************************

**SAQ 6.2c**

In one type of ELISA assay a limited amount of antibody is bound to a surface and enzyme-labelled antigen, and sample containing antigen is added. In an alternative type an excess of antibody is bound to the surface, the sample is added and this followed by an excess of enzyme-labelled antibody to generate a signal.

Which of the above two systems is non-competitive?

**Response**

The latter is a sequential, non-competitive type of assay which will generate a sandwich complex of the type:

$$\text{Support-Ab-Ag}_{(s)}\text{-Ab-E}$$

Reference to Fig. 6.2i might be helpful.

************************************

**SAQ 6.2d**

In a typical immunoenzymetric assay three steps are involved in which a sequence of attachments occur. Describe or illustrate diagrammatically these three attachment steps. Is this a competitive or non-competitive assay?

**Response**

A typical procedure for an ELISA assay using an enzyme labelled antibody would be as follows.

(*i*) Antigen (standard or sample) is added to a tube or well on which is adsorbed a large (excess) quantity of a specific antibody to give

Support -Ab-Ag

(*ii*) Excess enzyme labelled antibody (Ab-E) is now added and a sandwich is formed with antigen in the centre

Support -Ab-Ag-Ab-E

(*iii*) After removing surplus labelled antibody by decantation and rinsing, substrate is added to generate a signal.

Support -Ab-Ag-Ab-E-S

This assay is therefore non-competitive.

***********************************

**SAQ 6.2e**

Which assay method (EMIT or SLFIA) do you consider likely to have the lower detection limits? Explain the reasons for your choice.

**Response**

In EMIT assays an enzyme acts as label and will be measured by following the conversion of substrate to product. The catalytic nature of enzyme activity means that a continual change should occur giving a large and easily measured signal. The ability to distinguish this signal from background giving a high signal:noise ratio will contribute to a low detection limit. When a substrate is used as a label, it can only take part in one signal generating reaction (substrate to product conversion) and will produce a lower signal level.

***********************************

**SAQ 6.2f**

Illustrate the end product of a particle-based sandwich heterogeneous (separation) technique for the assay of an antigen using a fixed primary antibody.

Use the following symbols for the different components of the system:

support ⫴

antibody Ab

antigen Ag

particle labelled with antibody Ab-P

**Reponse**

The sequence of reaction is:

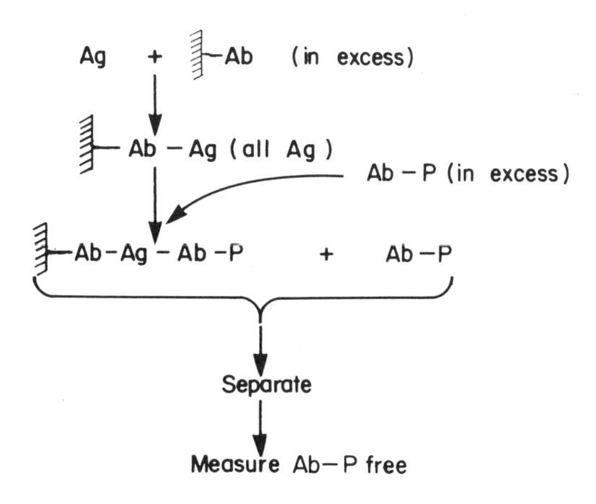

**SAQ 6.3a**

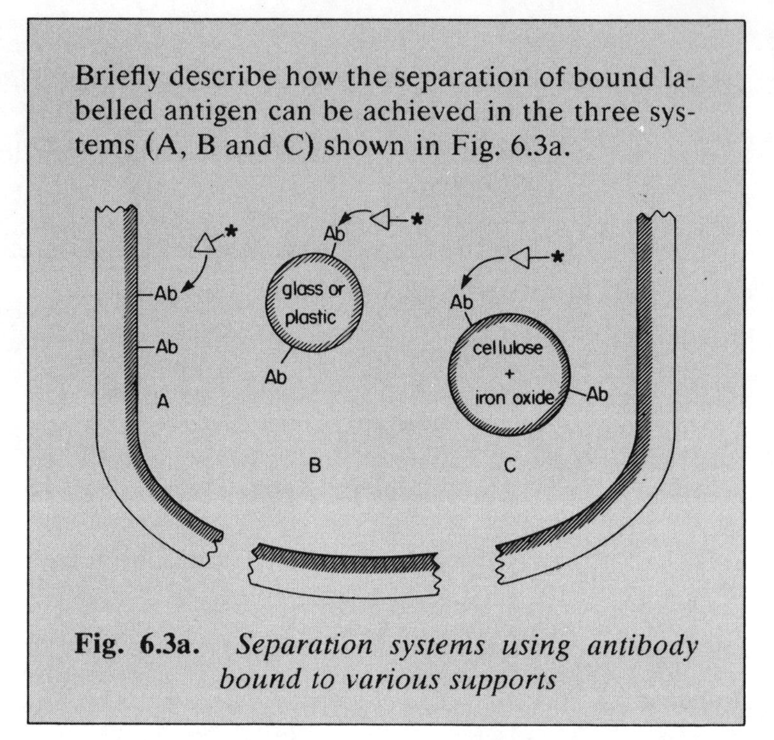

Briefly describe how the separation of bound labelled antigen can be achieved in the three systems (A, B and C) shown in Fig. 6.3a.



**Fig. 6.3a.** *Separation systems using antibody bound to various supports*

**Response**

A Decantation of unbound antigen remaining in the final phase.

B Collection of the particles carrying bound antigen by centrifugation or possibly spontaneous sedimentation.

C Collection of the particles carrying bound antigen by magnetic attraction of the iron core.

***********************************

**SAQ 6.5a**

Barbiturate drugs are commonly coupled to an antigenic carrier via their group on the C-5 position. If pentobarbitone (Fig. 6.5c) is used to generate an antibody, which of the compounds A (barbitone), B (phenobarbitone) or C (quinalbarbitone), do you think might cross-react with the resultant antibody?

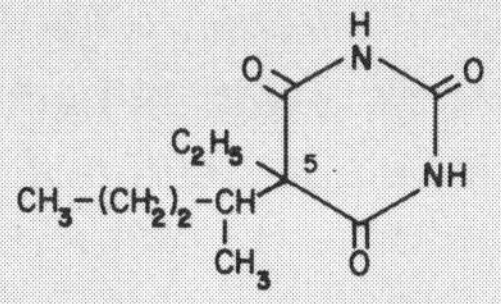

Pentobarbitone

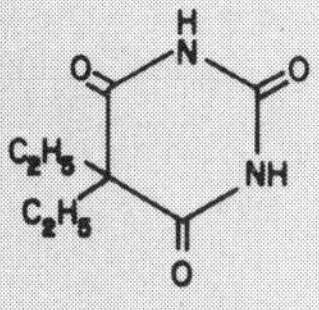

A Barbitone

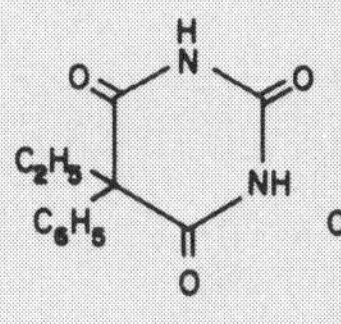

B Phenobarbitone

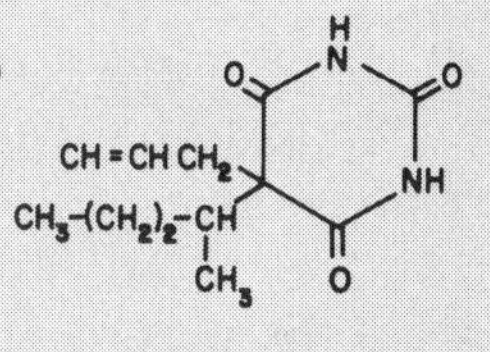

C Quinalbarbitone

**Fig. 6.5c.** *A comparison of the structures of some barbiturates*

**Response**

In practice significant cross-reactions with A and B occur due to the shielding effect of the coupling on the distinguishing substituents on C-5. The substituents on quinalbarbitone (C) are sufficiently large for a distinction still to be possible.

***********************************

**SAQ 6.5b**

For identical results between two assay methods, a correlation coefficient of 1 should be obtained, what should be the values of $m$ (the slope) and $c$ (the intercept)?

**Response**

$m = 1$, ie a given change in analyte concentration produces the same proportional change in signal in the two methods. The direct correlation gives a slope of 45° for identically spaced axes.

$c = 0$, ie at zero concentration of analyte both methods should give zero signal and therefore no intercept (positive or negative) on the $y$-axis.

***********************************

**SAQ 6.5c**

The drug diphenylhydantoin is of major significance in the treatment of epilepsy, particularly the 'grand mal' syndrome. Since the clinical symptoms of epilepsy are not easily monitored it is particularly important to be able to assay the serum concentrations frequently, easily and reliably. Many methods using different principles have been evolved and the following data show comparisons between

(*i*) a radioimmunoassay and an enzyme immunoassay method, and

(*ii*) the radioimmunoassay and a gas chromatographic method.

The correlation coefficient ($r$) can be calculated from the formula

$$r = \frac{\Sigma xy - \frac{(\Sigma x)(\Sigma y)}{N}}{\sqrt{\left(\Sigma x^2 - \frac{(\Sigma x)^2}{N}\right)\left(\Sigma y^2 - \frac{(\Sigma y)^2}{N}\right)}}$$

Where $x$ and $y$ are the two sets of data and $N$ the number of data points.

Calculate the correlation coefficient and comment on the results. Are there any other simple statistical investigations that could usefully be made? ⟶

**SAQ 6.5c (cont.)**

| Concentration of diphenylhydantoin (mg dm$^{-3}$) by: | | |
|---|---|---|
| Radio-immunoassay | Enzyme-immunoassay | Gas-liquid Chromatography |
| 5 | 7 | 7 |
| 7 | 9 | 10 |
| 6 | 8 | 8 |
| 8 | 11 | 12 |
| 9 | 11 | 11 |
| 5 | 6 | 7 |
| 10 | 13 | 13 |
| 6 | 8 | 12 |
| 8 | 9 | 9 |
| 9 | 12 | 12 |

**Response**

The equation for the calculation of $r$ looks complex but is in fact made up of a number of components which can easily be calculated independently. It is best therefore to draw out a table which includes these calculations and then to fit the results of these into the formula.

For ($i$)

| | $x$(RIA) | $y$(EIA) | $xy$ | $x^2$ | $y^2$ |
|---|---|---|---|---|---|
| | 5 | 7 | 35 | 25 | 49 |
| | 7 | 9 | 63 | 49 | 81 |
| | 6 | 8 | 48 | 36 | 64 |
| | 8 | 11 | 88 | 64 | 121 |
| | 9 | 11 | 99 | 81 | 121 |
| | 5 | 6 | 30 | 25 | 36 |
| | 10 | 13 | 130 | 100 | 169 |
| | 6 | 8 | 48 | 36 | 64 |
| | 8 | 9 | 72 | 64 | 81 |
| | 9 | 12 | 108 | 81 | 144 |
| $\Sigma$ | 73 | 94 | 721 | 561 | 930 |

and $N = 10$

$$r = \frac{721 - \frac{73 \times 94}{10}}{\sqrt{\left(561 - \frac{73^2}{10}\right)\left(930 - \frac{94^2}{10}\right)}}$$

$$= 0.964$$

Similarly for (*ii*)

| | $x$(RIA) | $y$(GLC) | $xy$ | $x^2$ | $y^2$ |
|---|---|---|---|---|---|
| | 5 | 7 | 35 | 25 | 49 |
| | 7 | 10 | 70 | 49 | 100 |
| | 6 | 8 | 48 | 36 | 64 |
| | 8 | 12 | 96 | 64 | 144 |
| | 9 | 11 | 99 | 81 | 121 |
| | 5 | 7 | 35 | 25 | 49 |
| | 10 | 13 | 130 | 100 | 169 |
| | 6 | 12 | 72 | 36 | 144 |
| | 8 | 9 | 72 | 64 | 81 |
| | 9 | 12 | 108 | 81 | 144 |
| $\Sigma$ | 73 | 101 | 765 | 561 | 1065 |

$$r = \frac{765 - \frac{(73)(101)}{10}}{\sqrt{\left(561 - \frac{73^2}{10}\right)\left(1065 - \frac{101^2}{10}\right)}}$$

$$= 0.780$$

These values show that the correlation between the RIA and GLC methods is worse than between the two immunoassay methods. This is not necessarily a serious problem if only one method is to be used and it is properly calibrated. However if you were to calculate the regression line you would see that the slopes of the relationships differ, but more importantly the comparison between RIA and GLC shows a significant intercept on the $y$-axis suggestive of the presence of some factor causing a constant elevation of the GLC response.

The equations for carrying out the calculation for slope ($m$) and intercept ($c$) of the straight line equation ($y = mx + c$) are

$$m = \frac{N\Sigma xy - (\Sigma x)(\Sigma y)}{N\Sigma x^2 - (\Sigma x)^2}$$

$$c = \frac{\Sigma y}{N} - m\frac{\Sigma x}{N}$$

The values obtained are

| | | |
|---|---|---|
| RIA versus EIA | $m = 1.238$ | $c = 0.359$ |
| RIA versus GLC | $m = 0.986$ | $c = 2.904$ |

**********************************

**SAQ 6.5d**

The experimental situation that forms the basis of this SAQ illustrates a common problem in the performance of ELISA and similar assays in which the binding of reagents to the plastic walls of microtitre plates is involved and serves to highlight the need to pay attention to the quality and reproducibility of apparatus, reagents etc in these assays. It is essential that the plastic should have consistent binding properties particularly across each plate but also ideally between plates.

The following data show the absorbance values of the contents of the wells of a typical plate of 12 columns and 8 (A–H) rows. While there are sophisticated statistical techniques for the comparison of such data can you make a subjective judgement as to whether any region of the plate is giving atypical results for the supposedly 96 replicate assays. Mean value is 1.06

| | 1 | 2 | 3 | 4 | 5 | 6 | 7 | 8 | 9 | 10 | 11 | 12 |
|---|---|---|---|---|---|---|---|---|---|---|---|---|
| A | 0.86 | 0.89 | 0.87 | 0.94 | 0.99 | 1.00 | 0.94 | 0.95 | 0.95 | 0.99 | 0.99 | 0.83 |
| B | 0.86 | 0.90 | 0.97 | 1.03 | 1.15 | 1.26 | 1.03 | 1.00 | 0.99 | 1.00 | 0.88 | 0.85 |
| C | 0.66 | 0.90 | 1.06 | 1.14 | 1.26 | 1.29 | 1.20 | 1.10 | 1.10 | 1.01 | 0.90 | 0.84 |
| D | 0.84 | 1.00 | 1.18 | 1.31 | 1.43 | 1.50 | 1.34 | 1.50 | 1.22 | 1.20 | 0.90 | 0.82 |
| E | 0.86 | 1.01 | 1.28 | 1.37 | 1.52 | 1.36 | 1.36 | 1.38 | 1.32 | 1.13 | 0.89 | 0.83 |
| F | 0.85 | 0.86 | 1.13 | 1.28 | 1.34 | 1.35 | 1.29 | 1.34 | 1.27 | 1.08 | 0.85 | 0.80 |
| G | 0.78 | 0.93 | 1.02 | 1.17 | 1.10 | 1.25 | 1.24 | 1.24 | 1.18 | 1.06 | 0.90 | 0.81 |
| H | 0.75 | 0.78 | 0.92 | 0.94 | 1.07 | 1.18 | 1.21 | 1.18 | 1.04 | 0.98 | 0.90 | 0.76 |

**Response**

The mean value for the absorbance values is 1.06 and a study of the distribution ought to show a tendency for higher values to occur in the lower centre of the plate. These values are identified in the diagram by italics.

| | 1 | 2 | 3 | 4 | 5 | 6 | 7 | 8 | 9 | 10 | 11 | 12 |
|---|---|---|---|---|---|---|---|---|---|---|---|---|
| A | 0.86 | 0.89 | 0.87 | 0.94 | 0.99 | 1.00 | 0.94 | 0.95 | 0.95 | 0.99 | 0.99 | 0.83 |
| B | 0.86 | 0.90 | 0.97 | 1.03 | ***1.15*** | ***1.26*** | 1.03 | 1.00 | 0.99 | 1.00 | 0.88 | 0.85 |
| C | 0.66 | 0.90 | 1.06 | ***1.14*** | ***1.26*** | ***1.29*** | ***1.20*** | 1.10 | 1.10 | 1.01 | 0.90 | 0.84 |
| D | 0.84 | 1.00 | ***1.18*** | ***1.31*** | ***1.43*** | ***1.50*** | ***1.34*** | ***1.50*** | ***1.22*** | ***1.20*** | 0.90 | 0.82 |
| E | 0.86 | 1.01 | ***1.28*** | ***1.37*** | ***1.52*** | ***1.36*** | ***1.36*** | ***1.38*** | ***1.32*** | ***1.13*** | 0.89 | 0.83 |
| F | 0.85 | 0.86 | ***1.13*** | ***1.28*** | ***1.34*** | ***1.35*** | ***1.29*** | ***1.34*** | ***1.27*** | 1.08 | 0.85 | 0.80 |
| G | 0.78 | 0.93 | 1.02 | ***1.17*** | 1.10 | ***1.25*** | ***1.24*** | ***1.24*** | ***1.18*** | 1.06 | 0.90 | 0.81 |
| H | 0.75 | 0.78 | 0.92 | 0.94 | 1.07 | ***1.18*** | ***1.21*** | ***1.18*** | 1.04 | 0.98 | 0.90 | 0.76 |

***********************************

# Units of Measurement

For historic reasons a number of different units of measurement have evolved to express quantity of the same thing. In the 1960s, many international scientific bodies recommended the standardisation of names and symbols and the adoption universally of a coherent set of units—the SI units (Système Internationale d'Unités)—based on the definition of five basic units: metre (m); kilogram (kg); second (s); ampere (A); mole (mol); and candela (cd).

The earlier literature references and some of the older text books, naturally use the older units. Even now many practicing scientists have not adopted the SI unit as their working unit. It is therefore necessary to know of the older units and be able to interconvert with SI units.

In this series of texts SI units are used as standard practice. However in areas of activity where their use has not become general practice, eg biologically based laboratories, the earlier defined units are used. This is explained in the study guide to each unit.

Table 1 shows some symbols and abbreviations commonly used in analytical chemistry. Table 2 shows some of the alternative methods for expressing the values of physical quantities and the relationship to the value in SI units.

More details and definition of other units may be found in the *Manual of Symbols and Terminology for Physicochemical Quantities and Units*, Whiffen, 1979, Pergamon Press.

**Table 1** *Symbols and Abbreviations Commonly used in Analytical Chemistry*

| | |
|---|---|
| Å | Angstrom |
| $A_r(X)$ | relative atomic mass of X |
| A | ampere |
| $E$ or $U$ | energy |
| $G$ | Gibbs free energy (function) |
| $H$ | enthalpy |
| J | joule |
| K | kelvin (273.15 + $t$ °C) |
| $K$ | equilibrium constant (with subscripts p, c, therm etc.) |
| $K_a$,$K_b$ | acid and base ionisation constants |
| $M_r(X)$ | relative molecular mass of X |
| N | newton (SI unit of force) |
| $P$ | total pressure |
| $s$ | standard deviation |
| $T$ | temperature/K |
| $V$ | volume |
| V | volt ($J\ A^{-1}\ s^{-1}$) |
| $a$, $a$(A) | activity, activity of A |
| $c$ | concentration/ mol $dm^{-3}$ |
| e | electron |
| g | gramme |
| $i$ | current |
| s | second |
| $t$ | temperature / °C |
| | |
| bp | boiling point |
| fp | freezing point |
| mp | melting point |
| ≈ | approximately equal to |
| < | less than |
| > | greater than |
| e, exp($x$) | exponential of $x$ |
| ln $x$ | natural logarithm of $x$; ln $x$ = 2.303 log $x$ |
| log $x$ | common logarithm of $x$ to base 10 |

**Table 2** *Alternative Methods of Expressing Various Physical Quantities*

1. **Mass (SI unit : kg)**

$$g = 10^{-3}\ kg$$
$$mg = 10^{-3}\ g = 10^{-6}\ kg$$
$$\mu g = 10^{-6}\ g = 10^{-9}\ kg$$

2. **Length (SI unit : m)**

$$cm = 10^{-2}\ m$$
$$Å = 10^{-10}\ m$$
$$nm = 10^{-9}\ m = 10Å$$
$$pm = 10^{-12}\ m = 10^{-2}\ Å$$

3. **Volume (SI unit : m$^3$)**

$$l = dm^3 = 10^{-3}\ m^3$$
$$ml = cm^3 = 10^{-6}\ m^3$$
$$\mu l = 10^{-3}\ cm^3$$

4. **Concentration (SI units : mol m$^{-3}$)**

$$M = mol\ l^{-1} = mol\ dm^{-3} = 10^3\ mol\ m^{-3}$$
$$mg\ l^{-1} = \mu g\ cm^{-3} = ppm = 10^{-3}\ g\ dm^{-3}$$
$$\mu g\ g^{-1} = ppm = 10^{-6}\ g\ g^{-1}$$
$$ng\ cm^{-3} = 10^{-6}\ g\ dm^{-3}$$
$$ng\ dm^{-3} = pg\ cm^{-3}$$
$$pg\ g^{-1} = ppb = 10^{-12}\ g\ g^{-1}$$
$$mg\% = 10^{-2}\ g\ dm^{-3}$$
$$\mu g\% = 10^{-5}\ g\ dm^{-3}$$

5. **Pressure (SI unit : N m$^{-2}$ = kg m$^{-1}$ s$^{-2}$)**

$$Pa = Nm^{-2}$$
$$atmos = 101\ 325\ N\ m^{-2}$$
$$bar = 10^5\ N\ m^{-2}$$
$$torr = mmHg = 133.322\ N\ m^{-2}$$

6. **Energy (SI unit : J = kg m$^2$ s$^{-2}$)**

$$cal = 4.184\ J$$
$$erg = 10^{-7}\ J$$
$$eV = 1.602 \times 10^{-19}\ J$$

**Table 3** *Prefixes for SI Units*

| Fraction | Prefix | Symbol |
|---|---|---|
| $10^{-1}$ | deci | d |
| $10^{-2}$ | centi | c |
| $10^{-3}$ | milli | m |
| $10^{-6}$ | micro | $\mu$ |
| $10^{-9}$ | nano | n |
| $10^{-12}$ | pico | p |
| $10^{-15}$ | femto | f |
| $10^{-18}$ | atto | a |

| Multiple | Prefix | Symbol |
|---|---|---|
| 10 | deka | da |
| $10^{2}$ | hecto | h |
| $10^{3}$ | kilo | k |
| $10^{6}$ | mega | M |
| $10^{9}$ | giga | G |
| $10^{12}$ | tera | T |
| $10^{15}$ | peta | P |
| $10^{18}$ | exa | E |

**Table 4** *Recommended Values of Physical Constants*

| Physical constant | Symbol | Value |
|---|---|---|
| acceleration due to gravity | $g$ | $9.81\ \mathrm{m\ s^{-2}}$ |
| Avogadro constant | $N_A$ | $6.022\ 05 \times 10^{23}\ \mathrm{mol^{-1}}$ |
| Boltzmann constant | $k$ | $1.380\ 66 \times 10^{-23}\ \mathrm{J\ K^{-1}}$ |
| charge to mass ratio | $e/m$ | $1.758\ 796 \times 10^{11}\ \mathrm{C\ kg^{-1}}$ |
| electronic charge | $e$ | $1.602\ 19 \times 10^{-19}\ \mathrm{C}$ |
| Faraday constant | $F$ | $9.648\ 46 \times 10^{4}\ \mathrm{C\ mol^{-1}}$ |
| gas constant | $R$ | $8.314\ \mathrm{J\ K^{-1}\ mol^{-1}}$ |
| 'ice-point' temperature | $T_{ice}$ | 273.150 K exactly |
| molar volume of ideal gas (stp) | $\mathrm{V_m}$ | $2.241\ 38 \times 10^{-2}\ \mathrm{m^3\ mol^{-1}}$ |
| permittivity of a vacuum | $\epsilon_0$ | $8.854\ 188 \times 10^{-12}\ \mathrm{kg^{-1}\ m^{-3}\ s^4\ A^2\ (F\ m^{-1})}$ |
| Planck constant | $h$ | $6.626\ 2 \times 10^{-34}\ \mathrm{J\ s}$ |
| standard atmosphere pressure | $p$ | $101\ 325\ \mathrm{N\ m^{-2}}$ exactly |
| atomic mass unit | $\mathrm{m_u}$ | $1.660\ 566 \times 10^{-27}\ \mathrm{kg}$ |
| speed of light in a vacuum | $c$ | $2.997\ 925 \times 10^{8}\ \mathrm{m\ s^{-1}}$ |

# Glossary of Terms

adjuvant

Substance or mixture which enhances the immune response to an immunogen in a non-specific fashion.

affinity

The strength of binding between an antigenic determinant or hapten and the corresponding binding site on an antibody.

agar

A heteropolysaccharide which is extracted from certain types of seaweed. It is used as a gelling agent in microbiological culture media.

agglutination methods

Methods which detect antibody-antigen reactions by the clumping of cells cross-linked in the process.

analyte

The particular substance that an assay is designed to measure.

antibiotic

A chemical substance produced by a microorganism or fungi that has the capacity to kill or inhibit other microorganisms while causing little or preferably no harm to the animal host.

antibody

Protein formed in response to exposure to an antigen and capable of binding with it specifically.

antigen

Substance capable of inducing the formation of antibody.

antiserum

Serum containing antibody of a defined specificity, ie from an animal immunised with the corresponding antigen.

avidity

Refers to the strength of binding between polyclonal antibody, eg in antiserum, and antigen or hapten.

B cells/lymphocytes

The lymphocytes principally responsible for antibody formation.

binary fission

In the context of microbiology binary fission describes the reproductive process which involves the division of a parent cell to give two daughter cells.

carrier (protein)

An antigenic molecule to which hapten is covalently linked. May also mean a naturally occurring blood protein with a carrier function.

chorionic gonadotrophin

A hormone which is responsible for maintaining the corpus luteum during the early stages of pregnancy and hence ensures the continued and increased secretion of progesterone before the placenta forms and can take over this function. The detection of this hormone in urine provides a means for the early diagnosis of pregnancy.

clone

A group of genetically identical individuals (cells or organisms) derived from a common parent by asexual reproduction, ie without any genetic variation.

competitive method

Two categories of molecules which are not distinguished by a binding substance bind randomly and indiscriminately to the latter. If the number of binding sites is limited, a competitive situation develops.

competitive protein binding assay, (CPBA)

A labelled ligand assay in which the binding agent is not antibody but either a transport protein or a cell receptor.

complement

A group of proteins permanently present in serum which are not induced and not specific to antigen but which may bind to antibody-antigen complexes either particulate or in solution (complement fixation).

complement fixation methods

Methods which detect antibody-antigen reactions by the rupture of cells which results when complement binds to antibody-antigen complexes formed on cell surfaces.

constant region (of immunoglobulin polypeptide chains)

In comparisons of different immunoglobulin molecules, the term refers to sections of a polypeptide chain with a relatively unvarying amino acid sequence (primary structure). Variation in the constant region (of the heavy chains) accounts for the different Ig fractions and their different physiological roles.

cross reaction

The same antibody able to bind (usually with differing affinities) different (but usually closely related) ligands.

determinant, (epitope)

An individual binding site on the antigen surface with which antibody can combine.

enzyme immunoassay, (EIA)

A labelled ligand assay in which antibody is the binder and the label is an enzyme. EMIT is a non-separation EIA and the labelled ligand variety of ELISA is a separation EIA.

enzyme linked immunosorbent assay, (ELISA)

A labelled ligand or labelled antibody assay which uses antibody as binder, an enzyme label and attachment to a solid phase to effect separation of free and bound label.

enzyme multiplied immunoassay technique, (EMIT)

A non-separation labelled ligand assay which uses antibody as binder and an enzyme label, which is inactivated when the label is in the bound state.

epitope

An individual binding site on the antigen surface, ie an antigenic determinant.

eukaryotic organisms

Organisms whose cells possess clearly defined nuclei containing matched pairs of chromosomes aside from their gametes (sex cells).

excess reagent method

A labelled binding assay which has the binding reagent in excess relative to the ligand, ie a non-competitive assay.

exotoxins

Exotoxins are responsible for severe or even fatal consequences of certain bacterial infections including diphtheria, tetanus and botulism. The exotoxin is released from the living bacteria and then diffuses away from the localised infection site to cause damage in other tissues of the body.

fluoroimmunoassay, (FIA)

A labelled ligand assay using antibody as binder and a fluorescent label. May be separation or non-separation (if the fluorescence of the labelled tracer in the bound state is sufficiently enhanced or quenched).

hapten

A small molecule which is not itself antigenic but which when combined with an antigenic carrier can elicit the formation of antibody that will bind the hapten alone.

heterogeneous method

Labelled binding assay which requires the separation of free and bound labelled tracer, ie a separation method.

heterologous antibody

Antibody of a different species of animal, ie antibody formed by one species against antigen from another.

histamine

An amine produced by the decarboxylation of histidine and found in all body tissues. It has many biological effects which include inducing increased permeability of the blood capillaries, lowering the blood pressure, contraction of most smooth muscle, increasing gastric secretion and accelerating the heart rate. Excessive release of histamine is associated with allergies and responsible for many of the symptoms seen in these diseases.

homogeneous method

A labelled binding assay which does not require separation of free and bound labelled tracer, ie a non-separation method.

immune complex

Product of an antibody-antigen reaction (may also include bound complement).

immune paralysis

Failure of the immune system to respond to large doses of antigen.

immune system

The total biochemical and cellular system responsible for the immune response.

immunisation (artificial)

Experimental or prophylactic exposure to an immunogen, usually by injection, ie prophylactic immunisation ≡vaccination.

immunity

Exemption from the symptoms of a disease, eg by prior immunisation.

immunoassay, (IA)

A general term for any method involving an immunological reaction, or sometimes specifically those in which the ligand is labelled.

immunoenzymometric assay, (IEMA)

A labelled antibody assay in which the label is enzymic, eg the labelled antibody version of ELISA.

immunofluorometric assay, (IFMA)

A labelled antibody assay in which the label is fluorescent.

immunogen

A substance capable of inducing an immune response whether humoral, cell mediated or both.

immunoglobulin

Antibody.

immunometric assay, (IMA)

An assay method using a labelled antibody.

immunoradiometric assay, (IRMA)

Labelled antibody assay using a radioisotope label.

immunosorbent/immunoadsorbent

Reagent antibody or antigen bound to a solid phase.

infectious disease

Disease caused by microbes, eg bacteria, viruses.

insulin

A hormone produced in the pancreas and which influences the utilisation of carbohydrates by the body. It is usually a deficiency of insulin which accounts for the symptoms shown by a patient with the disease diabetes mellitus.

latin squares

Statistically designed grid layouts which contain each component unit once in every row and column.

ligands

In this context, substances bound specifically by antibody, transport protein or cell receptor; they do not necessarily cross-link the binder molecules.

limited reagent method

A labelled binding assay which uses insufficient binding agent for the amount of ligand present, ie a competitive method.

luminescence immunoassay, (LIA)

A labelled ligand assay employing antibody as binder and a luminescent label.

lymphocytes

The cells primarily responsible for the immune response.

monoclonal antibody

Antibody formed by a single lymphocyte clone specific to a single (type of) determinant or hapten, effectively a single molecular species.

monospecific antibody

Polyclonal antibody purified from antiserum , eg by extraction using affinity chromatography with the appropriate ligand as the stationary phase.

nephelometer

An instrument designed to measure the light scattered by a sample which contains suspended particles.

non-self

Material foreign to the individual.

non-separation method

A labelled binding assay which does not require separation of free and bound labelled tracer, ie one in which the signal is significantly modified by the binding state of the tracer. A homogeneous assay.

pituitary gland

A small gland which is located just below the brain. It releases hormones into the bloodstream which exert stimulating or inhibiting effects on a wide range of endocrine glands and other tissues of the body.

polarization fluoroimmunoassay, (PFIA or PIA)

A non-separation immunoassay in which the fluorescent label shows a greater emission of plane polarized radiation when the label is bound to antibody.

polyclonal antibody

Antibody (in antiserum) formed in response to an antigen. A mixed population of immunoglobulin molecules produced by several different lymphocyte clones with a range of specificities and affinities, collectively specific to the different determinants on the same antigen.

precipitation method

Method of observing the reaction between a soluble antigen and antibody by the precipitation of large molecular aggregates presumably formed after extensive crosslinking.

prokaryotic organisms

Organisms such as bacteria and the blue-green algae which have no defined nuclear membrane and single, unpaired chromosomes.

prothrombin

A biologically-active substance which is involved in the blood clotting process. It is formed from an inactive, precursor substance called thrombin.

quantal response

A response which either does or does not occur, ie positive/negative, on/off etc.

radioimmunoassay, (RIA)

Labelled ligand assay using antibody as binder and a radioactive isotope as label.

receptor assay

Labelled ligand assay using a cell receptor as binder.

saturation analysis

Analysis by a competitive method, eg a labelled ligand assay in which the limited number of binder molecules become saturated with ligand.

secondary antibody

Antibody raised in one species of animal against the antibody of another, ie heterologous anti-immunoglobulin antibody.

self

Constituent of the individual.

self-antigen

Constituent of an individual which in other species or other individuals of the same species can induce antibody formation.

separation method

A labelled binding assay which requires a physical separation of the free and bound labelled tracer before measurement, ie a heterogeneous assay.

substrate labelled fluoroimmunoassay, (SLFIA)

A non-separation immunoassay employing a substrate label whose enzyme-catalysed conversion to a fluorescent product is blocked when the tracer is bound to antibody.

temporal variation

Variation which is related to the time over which an activity is performed.

time resolved fluoroimmunoassay/phosphoroimmunoassay, (TR-FIA).

An immunoassay using a fluorescent or phosphorescent label whose signal is measured after a short time span following the end of the period of excitation.

titre

Reciprocal of the dilution of an antibody preparation which binds 50% of a given amount of ligand. It is a function both of the concentration and affinity of the antibody.

tumour specific antigens

Antigens characteristic to a tumour and which are absent from the normal cells of an individual.

two-site immunometric assay

A labelled antibody assay using both unlabelled antibody bound to a solid phase to effect separation, and labelled antibody to provide a signal. Usually the two antibodies are monoclonal, specific to different well separated determinants on the antigen.

valency (of a binding agent or ligand)

The number of binding sites per molecule.

variable region (of immunoglobulin polypeptide chains)

In comparisons of different immunoglobulin molecules, this term refers to sections of a polypeptide chain with a highly variable amino acid sequence. Responsible for antigen specificity and forming the sites of antigen binding.